W9-BHU-316

NOV 2 0 1998

Science Library

FRACTALS
IN CHEMISTRY

FRACTALS IN CHEMISTRY

WALTER G. ROTHSCHILD
Department of Chemical Engineering and Materials Science
Wayne State University
Detroit, Michigan

A Wiley-Interscience Publication
JOHN WILEY & SONS, INC.
New York • Chichester • Weinheim • Brisbane • Singapore • Toronto

This book is printed on acid-free paper. ∞

Copyright © 1998 by John Wiley & Sons, Inc. All rights reserved.

Published simultaneously in Canada.

No part of this publication may be reproduced, stored in a retrieval system or transmitted in any form or by any means, electronic, mechanical, photocopying, recording, scanning or otherwise, except as permitted under Sections 107 or 108 of the 1976 United States Copyright Act, without either the prior written permission of the Publisher, or authorization through payment of the appropriate per-copy fee to the Copyright Clearance Center, 222 Rosewood Drive, Danvers, MA 01923, (978) 750-8400, fax (978) 750-4744. Requests to the Publisher for permission should be addressed to the Permissions Department, John Wiley & Sons, Inc., 605 Third Avenue, New York, NY 10158-0012, (212) 850-6011, fax (212) 850-6008, E-Mail: PERMREQ@WILEY.COM.

Library of Congress Cataloging-in-Publication Data:

Rothschild, Walter G., 1924–
　　Fractals in chemistry / Walter G. Rothschild.
　　　　p.　cm.
　　"A Wiley-Interscience publication."
　　Includes bibliographical references and index.
　　ISBN 0-471-17968-X (cloth : alk. paper)
　　1. Chemistry, Physical and theoretical—Mathematics.　2. Fractals.
　I. Title.
　QD455.3.F73R67　1998
　541—dc21
　　　　　　　　　　　　　　　　　　　　98-10104
　　　　　　　　　　　　　　　　　　　　CIP

Printed in the United States of America

10 9 8 7 6 5 4 3 2 1

CONTENTS

PREFACE

This book addresses the researcher, practitioner, teacher, and advanced student in chemistry, bringing various but selected representative subjects together where use of fractals bestows advantages to the chemical sciences and their technology: Understanding fractals may provide novel insights, a general, more predictive, more useful classification scheme, or an explanation of a phenomenon where a classical approach had led to a dead end. In addition, seemingly divergent and otherwise different phenomena may have common attributes when examined from underlying fractal viewpoints. The field has now sufficiently matured so that critical views and opinions can be advanced, at least on some aspects of the use of fractals in chemistry.

The text is the result, in part, of teaching a course dedicated to fractals to graduate and interested students at Wayne State University, Detroit, Michigan, during several semesters. In addition, the book reflects 33 years of fundamental and applied research, predominantly in gaseous and condensed-phase spectroscopy of structure and dynamics, combustion phenomena, and automotive catalysis, at the Scientific Research Laboratories of the Ford Motor Company in Dearborn, Michigan. Furthermore, lengthy stays at the Laboratory for Molecular and Crystalline Spectroscopy of the University Bordeaux I (in Talence, France) greatly advanced the present purpose. It was quite unavoidable to fall onto fractals in studies of such strange phenomena and irregular objects as relaxation and aggregation dynamics of super-cooled liquid-crystal fluids on one hand, and of chemisorption characteristics of high-area adsorbents on the other. It eventually became obvious that an interplay of randomness, feedback effects, system constraints, and "variations on a theme," precisely the conditions leading to fractals, was ever-present.

I considered it prudent strategy to offer a bounded set of subjects: Trying to include everybody's interests on one hand, or to be too specialized on the other, pleases no one in the end. To avoid either extreme, subjects from the original literature were selected that responded to the bread and butter issues to chemists, namely, fractal aspects of surfaces and porosity, of aggregation phenomena of inorganic and organic materials during deposition and diffusion processes, chemical reactivity, turbulent premixed flames, and elastic and dispersive spectroscopies, just to name a few. The Contents lists more details; presentation of material from monographs devoted to a specific fractal topic was kept to a minimum.

The reader will find an appropriate treatment of fractons as well as some discussions of less common fractal aspects, for instance, of physicochemical properties of homologous compounds and ablation of food particles. But in general, the emphasis is on rather ordinary chemical systems and how they now appear to us, as well as how they may be regarded given consideration of their fractal aspects. Hopefully, this approach will encourage readers to further inquiry and reading.

The text is concise—discussion of the chemistry per se is kept short and serves mainly to refresh the reader's memory—with extensive intra- and intersection references to data, equations, discussions, notes, and glossaries. It relies heavily on many figures, almost all redesigned from their originals and often thoroughly modified and sharpened to fit present purpose and plotted using Sigma Plot® 4.0 for Windows®, Copyright © 1997 by SPSS Inc.

ACKNOWLEDGMENTS

I am grateful to Professor D. Carl Freeman of the Biological Sciences Department of Wayne State University, for support, encouragement, innumerable and lengthy discussions (some rather heated), and critical reading and suggestions to improve the manuscript in content and style. Furthermore, I am thanking Dr. Adolf E. de Vries of the FOM Institute in Amsterdam for his detailed comments, and my many students, particularly Mr. Ian W. Ladomer and Mr. Robert L. Mara, for their attention, participation, and helpful questions during the presentations of the course material.

I also would like to express my gratitude to the publishers acknowledged in the figure captions for their permission to reprint or use copyrighted material.

I owe the suggestion to write this book as well as continuing interest to the late Dr. Rita G. Lerner, Editor, Consultant, colleague, and friend since my graduate student days at the Chemistry Department of Columbia University in New York City.

1

THEORETICAL OUTLINE

1.1 INTRODUCTION

Most chemists have heard about fractals; *Chemical Abstracts* lists the item under their proper heading for some years now.

With all the books, journals, and reviews devoted to fractals, why write a comprehensive text addressed to chemists? Most of the extant books do not seem to think of the chemist as a customer: Their content is simply too mathematical, often concentrating on the deeper, novel, controversial, or esoteric points of fractal *theory*. Less is offered in the way of information on wider chemical applicability and utility. Thus, the following discussions are planned and designed to make a stronger case for the *use* of fractals in chemistry: Experimental results, from the literature of wide-ranging fractal applications to chemistry, are discussed. Furthermore, as a long-time practicing chemist, I address how ideas from a fractal viewpoint can be useful in chemical applications and understanding where—as far as I know—none are published. I am pretty sure all this will convince some of my readers to apply fractals, as well as to explore them, to a larger extent than before, having better appreciated what fractals can and cannot do.

Discussion of experimental fractal aspects in chemistry without including the underlying fractal theory is futile. Fractal theory is not difficult but rather unfamiliar, dealing with sets, measures, topologies, and so on, topics a chemist does not ordinarily learn in graduate school [1]. Thus, only the minimum of theory needed to appreciate the content is presented in text and notes. For those interested in eventually going further, more detailed discussions of some fundamental topics, highlighted *italicized–underlined* at their first appearance, are found in Glossaries 1 and 2 at the end of the text.

Still, a chemist may wonder "Why bother about fractals; my work has been going well for years, thank you." Perhaps it is of interest to relate how I became involved. My research topics included nothing extraordinary: Criticality of metastable phases, relaxation dynamics in amorphous media, rates of catalyzed oxidation–reduction steps, distributions of chemisorption sites on supported metals, self-ignition dynamics of compressed air–fuel mixtures, hydrocarbon speciation of engine effluents. Seemingly well-defined objectives approachable by well-delineated methods. Yet, there was frequently divergence between experimental results and theory that could not simply be excused as "noise." Furthermore, it was often difficult to precisely fix initial conditions, or even define them well.

By accident, during a university visit, I opened a book with beautiful designs and strange geometries that described a new way of looking at complex structures and dynamics: Mandelbrot's *The Fractal Geometry of Nature* [2]. The major promise of the approach, it seemed to me, was that a view of local conditions implied the view of the whole. Complex structures could be generated by iterations of rather simple codes—perfect for some initial computer probing. Desired degrees of randomness could be readily incorporated and it was no longer necessary to fix all the system parameters.

As always, a price had to be paid: An important fractal signifier of the whole, the fractal dimension, is silent on the detailed structure and absolute size of the object. Let us begin with a simple illustration.

1.2   AN EXAMPLE: FRACTAL SURFACE MODEL
AND ITS FRACTAL DIMENSION

We are going to explore the generation of increased surface area, advantageous in adsorption and catalysis, by a computer model and compare the outcome with laboratory experiments. Such an approach not only introduces several important and ever occurring fractal aspects of interest to chemists but is also quite relevant and ubiquitous.

The surface area augmentation process may be pictured as erosion by electrolysis or, simpler, pitting by sandblasting. As the object, we may consider an originally smooth surface, say a thick Pt foil. We introduce probability concepts because we are ignorant of impact strengths, angles of incidence, frequency and location of the pitting events. Assumptions made are: (a) These events are independent, caused by impinging individual, noninteracting "sand particles." (b) The events occur at random locations on the surface. (c) A single pitting event is realized by a small surface indentation of constant, predescribed size. (d) The process is iterated event by event; a large number of events are run.

All this is readily graphed [3] in two dimensions on the screen of the monitor by considering the *profile* of the generated surface indentations: Figure 1.1(*a*) shows two profiles of 2000 iterations, each at different runs of the random number generator but identical "pit shape" of 2(depth) $\times$ 4 (width) relative to a screen width of 640. Consider these results as Y–X maps, or graphs, of coordinate points (Y, X) of indentation depths Y versus abscissa positions X—in other words, the profile (or contour) of the surface.

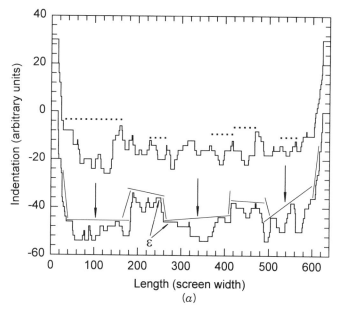

Fig. 1.1(a)   Profiles resulting from the surface-pitting model by iterative random impinge-ments of rectangles of 2 (depth) $\times$ 4 (width) in units of the abscissa length. Ordinate = height below the original line (positioned at 0). Abscissa = position on profile. Two separate runs of 2000 pitting events each are shown, vertically displaced by 30 units to avoid crowd-ing of the figure. Examples of regions of self-similar overall shape are designated by over-strike dotted lines (upper portion of the figure). Two different linear elements ε (yardstick) are indicated; vertical arrows point to the centers of three large indentations of $\varepsilon \approx 150$ each (lower portion of the figure). Figure 1.1(b) on next page.

First, on the basis of intuition, the result in Fig. 1.1(a) is not at all what one would have expected to see, namely, more or less uniformly distributed indentations of the order of the size of the impinging particle. But this is not what Fig. 1.1(a) displays. Looking at the plot from a distance, a few singularly wide indentations appear. Looking closer, we discern a great number of small, underlaid by a greater number of smaller, indentations.

Second, the plots show a *symmetry* although the process is random: Different re-gions of a profile—marked by overlaid point lines in the upper portion of Fig. 1.1(a) for easier seeing—look alike if their *absolute* size is ignored. This outcome is an example of (statistical) *self-similarity*.

Third, there is an approach to a limit structure or *attractor:* Further iterations of the code may well lead to a different overall appearance, but they do not materially change the *number distribution* of the various pit diameters. There will always be the same number of large pits within the screen width of the monitor, on average, under different values of the starting point (seed) of the pseudorandom number generator.

Let us discuss further the curious relation between the number $\mathcal{N}(\varepsilon)$ of pits and the pit size ε (diameter) noticed on the profiles [Fig. 1.1(a)], namely, that there are many more smaller than larger indentations. Note, for instance, three relatively huge pits of wall-to-wall distance $\varepsilon \approx 150$ and of net depth $Y \approx 13$, their respective centers marked by arrows. Solely in terms of these three pits, roughly estimate length $L_{\rm C}$ of the profile as $3 \times 150 + 6 \times 13 + 2 \times 90 \approx 700$—a factor of about 1.2 over initial length 640.

Now inspect Fig. 1.1(b), which displays a finer measuring scale of $\varepsilon \approx 50$. Find $L_{\rm C} = 31 \times 50 = 1550$. This value is already a factor of 2.6 over the original length 640. Hence, it seems that a measurement of the length of the profile depends on the size of the yardstick applied!

For precise results, this empirical counting method would have to undergo refinement. First, several computer runs are needed for a good average. Second, the number of pits, $\mathcal{N}(\varepsilon)$, has to be counted over the entire accessible range of ε. However, for present purposes there exists a more direct way that also introduces fundamental concepts frequently used to characterize fractals generated by processes containing aspects of randomness. To appreciate this particular approach, consider the kernel of the pitting model, namely, iterative generation of small, equal-size indentations Y at unrestricted, uniformly distributed random abscissa positions X. In principle, it should not matter where the random process takes place on the surface profile. Furthermore, it also cannot make a difference how we sequence plotting of the Y–X data of the indentations because any indentation step is uncorrelated with any other: We may instantaneously graph depth coordinate Y at its abscissa position X in an event-by-event fashion—a video of (Y, X) flashing here and there on the screen. Or we may store the accumulating Y–X values in a file and show the photograph at the termination of the run.

It is not difficult to demonstrate that these scenarios are equivalent to the

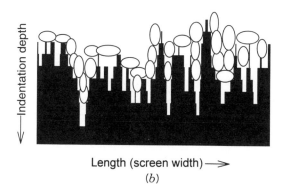

Length (screen width) $\longrightarrow$
(b)

Fig. 1.1(b) Schematic of counting the number of irregularities of a profile by ellipsoids of about 50 units on their long axis, relative to the screen width. The displayed object represents a partial image of a computer run that folds an initially straight line by iterated random indentation of relative width 0.014 and depth 0.0068 each [see also Fig. 1.1(a)]. This figure depicts the folded line after 1500 iterations.

"position–elapsed time" map of the motion of a particle performing a random walk on a line—stretching theory a bit to suit our need: Figure 1.2 explains this better than any words. The displayed example is an arbitrary, consecutive 14-step random walk, written RLLRLRRRLRLLLR (R: step right; L: left): Scheme (*a*) merely lists the assumed individual R or L jumps (of equal step size), commencing at (arbitrary) position 0; changes in direction are displaced vertically upward for better viewing. Scheme (*b*) shows the actually traversed path of the walker on its prescribed space (a line), namely, a total of net three jump lengths. Finally, scheme (*c*) displays the walker's momentary position (Y, X) with $X = time$ elapsed. Scheme (*c*) is the walker's Brownian record, graph, or map [2, 4]: Obviously, the Brownian map gives the clearest demonstration of the walker's progress. For instance, it indicates that the walker has revisited its original starting position 0 five times, position 1 four times, position 2 two times, position -1 three times, and so on, within the observed interval.

Hence, it would seem that we can generate the profile in yet another, more general way, namely, as a movie. (a) The computer "flips a coin." If heads or tails, coordinate point $(Y_0, X_0) = (0, 0)$ performs a (constant size) jump δY up or down, to $Y_0 \pm \delta Y = Y_1$, respectively. (b) Subsequently, Y_1 is moved by small (constant size)

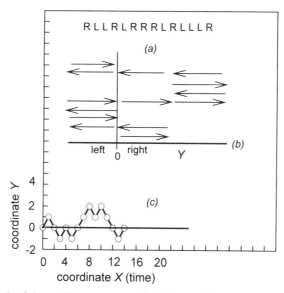

Fig. 1.2 Sketch of three different ways of depicting a (14-step) random walk. Conditions: The walker is restrained to a line (dimension 1); each single jump is of constant length and may go either left or right with equal probability. Part (*a*): Scheme of the walk under arbitrary starting point 0. Steps changing direction are vertically displaced (bottom to top) for better viewing. Part (*b*): Total path traversed by the random walker on its space. Part (*c*): Position–time record or map of the random walk.

positive increment δX to $X_0 + \delta X = X_1$. (c) Old and new positions, (Y_0, X_0) and (Y_1, X_1), are connected by a straight line [see Fig. 1.2(c)]. (d) Operations (a)–(c) are repeated with point (Y_1, X_1), and so on. The result may not look precisely like the profile of Fig. 1.1(a) because it uses the Euclidian distance "as the crow flies" and not a Manhattan _metric_ between adjacent points. However, this should not matter for small δY, δX.

Thereby, we have generated a jagged curve, meandering from left to right on the screen, that describes the Y–X map of Brownian random walk on a line. Now we have to find out how many pieces of linear size ε, the pit diameters, make up the length of the generated random profile—equivalent to the random surface-pitting model. From the theory of Brownian walk [5], we know that coordinates Y, X of the walker's map [see Fig. 1.2] are related by $\langle Y^2 \rangle^{1/2} \sim X^{1/2}$, where $\langle Y^2 \rangle^{1/2}$ is the (root-mean-square) distance between two succeeding (Y, X) in the limit of many random jumps (long X) [6]. Then, in order to obtain the cumulative distance between points (Y, X) progressing from left to right [Fig. 1.2(c)] as the walker runs by L, R random jumps [Fig. 1.2(a)] over its allotted space [Fig. 1.2(b)], we merely add all increments, $X^{1/2}$, $\Sigma X^{1/2} \approx \int X^{1/2} dX \sim X^{3/2} \approx \varepsilon^{3/2}$. Finally, we take the reciprocal of this result as the _number $\mathcal{N}(\varepsilon)$ of elements of linear size ε that make up an entire random profile_ [see Figs. 1.1(a) and 1.2(c)] in units of ε: $\mathcal{N}(\varepsilon) \sim \varepsilon^{-3/2}$. The smaller ε, the larger $\mathcal{N}(\varepsilon)$, evidently. (Purists may want to look already at [8] for a more precise scaling argument.)

How does this help us in understanding what a fractal is like? To see this, we replace the particular value $\frac{3}{2}$ of the exponent of ε by _general_ exponent d, writing

$$\mathcal{N}(\varepsilon) \sim 1/\varepsilon^d \sim \varepsilon^{-d} \tag{1.1}$$

Taking logarithms on both sides of Eq. (1.1), forming their ratio, and going to _ever_ smaller ε, then leads to $\log[\mathcal{N}(\varepsilon)]/\log(\varepsilon) = -d$ or $\log[\mathcal{N}(\varepsilon)]/\log(1/\varepsilon) = d$, $\varepsilon \Rightarrow 0$. Now, _if_ this ratio exists, exponent d turns out to be the _dimension_ of the object whose size distribution $\mathcal{N}(\varepsilon)$ we just counted [2].

Why would exponent d represent a dimension? Figure 1.3 demonstrates the principle for two well-familiar situations, namely, a line of length a and a rectangle of area $a \times b$. We proceed by marking the object into a number of $n = 2, 3, \ldots$ equal-sized smaller parts, each of the same shape as the original whole (self-similarity). Notice that their number $\mathcal{N}(\varepsilon)$ goes with ε as ε^{-1} for the line (one-dimensional) and ε^{-2} for the rectangle (two dimensional), whatever the degree of partition of the object into self-similar pieces of _linear_ extent $\varepsilon = a/n$ (the line) or $\varepsilon = (ab)^{1/2}/n$ (the rectangle). Extension to three dimensions is obvious.

We have simply performed (a) a _scaling_ operation and (b) a _covering_ or _measure_ of self-similar objects: In more general terms then, Eq. (1.1) performs the operations for any value d. Clearly, _exponent d of ε is an invariant quantity_: It does not depend on the value of "yard stick" ε but only on the production or algorithm of the process. Evidently, dimension d may be _non_integer. In fact, for our purposes it may assume any value between 0 and 3: $0 \leq d \leq 3$[1, 2] (see also _Hausdorff measure_ in Glossary 1). If so, this characterizes the object as a fractal. Obviously, length

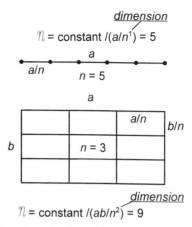

Fig. 1.3 *Upper part:* Partition of a line of length *a* into *n* = 5 self-similar pieces of linear size *a*/*n* each. *Lower part:* Partition of a rectangle of area *a* × *b* into n^2 = 9 self-similar pieces of *linear* extent $(ab)^{1/2}/n$ each.

$L_C \sim \varepsilon^{-d}\varepsilon \sim \varepsilon^{1-d}$ of the profile diverges at smaller and smaller ε as its dimension $d = 1.5$ turns out to be larger than that of its line elements ($d = 1$).

What does the dimension of such strange noninteger values *signify?* Just some exotic number? A most profitable way to look at fractal dimension in chemistry is to consider it a quantifier of an object's *capacity of filling available space*. (This concept will be extended later in the text.) For instance, the generated fractal profile of the fractal surface fills its available space—the two-dimensional plane of the paper where it is plotted—to a greater extent than the original straight line. The effective dimension of the profile must be larger than that of a line segment *if* the space-filling property occurs across wider *linear scales*. In theory infinitely wide—in our physical world over a sufficiently wide range [1, 2]. We shall see below how such complex geometries introduce novel concepts into chemistry. Consequently, a fractal dimension is not just another number but makes a profound statement on the property of the fractal object.

But, to return to some specifics concerning our surface-enlarging model: There is additional justification of drawing the parallel between random pitting and Brownian motion by investigating how the profile would look if scaled by some factor. Assume it were desirable to include a portion of Fig. 1.1 (*a*) into a report but that a reduction in its width by a factor ⅔ were required. Could we then take the figure to the copier down the hall and set it at magnification equal to 0.67? The answer is no: The map of the Brownian random walk is a *self-affine fractal,* meaning that ordinate *Y* and abscissa *X* are scaled each by a *different* constant factor. For the Brownian map of the profile, we have to scale the ordinate not by abscissa factor ⅔ ≈ 0.67 but by its square root $0.67^{1/2}$ ≈ 0.82 [7].

Whereas theoretical scaling arguments to rationalize this requirement are found in [8], Fig. 1.4 gives pictorial proof. The original object, placed at the bottom of

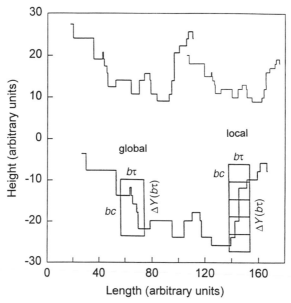

Fig. 1.4   Scaling in the surface-pitting model of Fig. 1.1(*a*). *Bottom part:* A portion of the profile, as generated by computer-modeled random surface pitting. The coverings of its self-affine patterns, $\Delta Y(b\tau)$, by small boxes of height bc and width $b\tau$ are sketched for high-resolution (local) and low-resolution (global) probing. *Upper portion, left:* Self-affine version of the bottom object scaled by a factor $b = \frac{2}{3}$ on the abscissa and by a factor $b^{1/2} = (\frac{2}{3})^{1/2} \approx 0.82$ on the ordinate. *Upper portion, right:* Version of the bottom object scaled isotropically by factor $b = 0.50$ on a copier.

Fig. 1.4, is a profile generated by the surface-pitting program. The upper left profile displays the results of scaling it by 0.67 on the abscissa, $0.67^{1/2} = 0.82$ on the ordinate, as mentioned just above. We must have done it right, because the *shapes* of original and down-scaled object are indistinguishable: They are indeed self-affine under arbitrary scale factor b on the abscissa and $b^{1/2}$ on the ordinate. Not so for the object placed at upper right in Fig. 1.4: It was obtained by reducing the original on a copier at a magnification factor of 0.50. Compared with the profile to its left, it is evidently stretched horizontally.

   Finally, to obtain the fractal dimension of the generated Pt *surface,* we imagine placing a large number of its profiles—each from a new computer run—next and parallel to each other, such as in a piece of pre-sliced smoked salmon; we effectively add a coordinate axis (dimension 1). Hence, $d(\text{surface}) = 1 + \frac{3}{2} = 2.5$. Figure 1.5 gives the general idea.

   Summarizing:

- The random iterative pitting process generates intricate patterns from a simple starting object. The pattern approaches a limit structure definable by some invariance criterion, namely, exponent d of the power law distribution of the

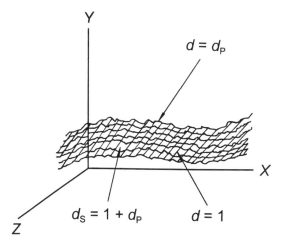

Fig. 1.5 The X–Y–Z view of the portion of an arbitrary surface of fractal dimension $d = 1 + d_p > 2$ as generated by adding coordinate axis Z of $d = 1$ to the Y–X profiles of common fractal profile dimension $d = d_p > 1$. The profile dimension amounts to $d_p = \frac{3}{2}$ for the random surface-pitting model of Fig. 1.1(a).

    number of discernible pieces making up the whole object, $\mathcal{N}(\varepsilon)$, with linear *size* ε of a piece [Eq. (1.1)].

- The surface generated in this computer experiment yields a distribution of irregularities at finer and finer detail within some physical range of ε [9]. Under decreasing linear size ε, irregular surface area increases at a faster and faster rate [10].

- The dimension of the generated profile, $d = 1.5$, or the dimension of its surface, $d = 1 + 1.5 = 2.5$, is larger than that of its constituent line elements, $d = 1$, or larger than that of its constituent surface elements, $d = 2$, respectively.

- The space-filling capacity of the surface fractal, expressed by the excess of its fractal dimension d above classical value $d = 2$, attains a higher occupancy of its maximally available three-dimensional space than the ordinary surface. Surface area $S_C \sim \mathcal{N}(\varepsilon)\varepsilon^2 \sim \varepsilon^{2-d}$ diverges for $\varepsilon \Rightarrow 0$ because the value of fractal dimension $d(\text{surface}) = 2.5$ exceeds 2.

- The limit structure of the profile (its trail within the Y–X plot) is a (statistically) self-similar fractal curve of dimension $\frac{3}{2}$. The limit structure of its Y–X plot (its record or map) is a (statistically) self-affine fractal of local dimension $\frac{3}{2}$ and global dimension 1.

- The process is of the nature of an iterative coordinate <u>*contraction-shift*</u> operation [11, 12], leading to ever increasing detail over some range: A smooth area (or its profile) is effectively crumpled into a set of many contiguous surface (or line) elements, their number $\mathcal{N}(\varepsilon)$ being distributed by a power law [Eq. (1.1)] over the range of linear quantity ε under invariant exponent d, the fractal dimension. Although only a model [13], some fundamental aspects of surface

fractality, fractal dimension, self-similarity, self-affinity, and randomness have
been introduced with its help without too much difficulty.

NOTES AND REFERENCES

[1] M. F. Barnsley, *Fractals Everywhere,* 2nd ed. (Academic, New York, 1993). This book
is relatively descriptive and accessible to chemists. It contains the solutions to instructive problems. For a text on set theory for self-study, see D. Goldrei, *Classic Set Theory*
(Chapman & Hall, New York, 1996).

[2] B. B. Mandelbrot, *The Fractal Geometry of Nature* (Freeman, New York, 1983). Not an
easy book, but it is a must on the fractal practitioner's bookshelf.

[3] After drawing a horizontal line over the screen—representing the profile of the virgin
foil (three-dimensional plotting not offering better understanding)—an iterative scheme
of single-pitting events is modeled as follows. The program picks a random number
from the clock of the computer, scales it to the effective width of the monitor screen,
and places, at that very X coordinate, a small (constant-size) indentation (Y coordinate).
At the termination of the process, all adjacent X–Y points are connected by straight
lines and the final profile appears on the screen and is written into a file. I am grateful
to Dr. Pierre M. Willermet, Ford Motor Company, Scientific Research Laboratories in
Dearborn, MI, for having furnished me with a copy of the program (in QBasic).

[4] J. Feder, *Fractals* (Plenum Press, New York, 1988), Chapter 9.2.

[5] Recall Einstein's formula for random walk, $\langle Y^2 \rangle \sim X$, hence $Y \sim X^{1/2}$, where Y is the
root-mean-square distance "as the crow flies" traveled by the walker during time inter-
val X.

[6] The tilde sign ($\sim$) indicates a functional relation or equivalence without worrying about
uninteresting proportionality constants. Symbol $\approx$ signifies that the precision of a numerical value is not further specified. Symbol $\equiv$ defines an assigned identity, usually
applied to simplify equations. The equality sign ($=$) links a numerical value to some
general parameter or indicates an equality between expressions.

[7] First, see [5] above: Scaling abscissa X by b, $X \Rightarrow bX$, requires scaling of ordinate
$Y \Rightarrow (bX)^{1/2} \sim b^{1/2}X^{1/2} \sim b^{1/2}Y(X)$. Consequently, $Y(X) \sim b^{-1/2}Y(bX)$. This operation is
an affine transformation (see also Chapter 9, Section 9.3 of [4]).

[8] A covering (measure) of the self-affine fractal object is attempted as follows (see bottom part of Fig. 1.4): We partition the total time span T of the record—its abscissa
range—into $T/b\tau$ segments of length $b\tau$ each, where b is a scale factor on the abscissa
(see [7]). Within each segment, the corresponding (root-mean-square) ordinate variation is then $\Delta Y(b\tau)$, where $b\tau$ designates the function argument. As sketched in the figure, $\Delta Y(b\tau)$ can be covered by a stack of boxes, each of width $b\tau$ and height bc, such
that, for a given τ and b, the cover is reasonably truthful to the shape of $\Delta Y(b\tau)$. In other
words, the number of boxes of the stack ought to parallel the number of irregularities
within $\Delta Y(b\tau)$, necessitating picking parameter c to be *sufficiently small*. (Note that the
stack to the left does not satisfy this criterion.) Then, the number of boxes covering
$\Delta Y(b\tau)$ over time span T is obtained via $\mathcal{N}(b; c, \tau)$. Hence, quite generally, $\mathcal{N}$ = area
(total)/area (box) = $(\Delta YT)/b\tau bc = (\Delta Y/bc)T/b\tau \sim b^{-d}$, the last relation in this sequence
arising from Eq. (1.1) with $\varepsilon \equiv b$. [The function argument of $\Delta Y(b\tau)$ is omitted here to
avoid confusion.]

We are now looking for the value of d under high (local) and low (global) resolution coverings, as shown in Fig. 1.4. *Global:* $bc = \Delta Y$, thus $(\Delta Y/bc)T/b\tau = T/b\tau$. Then, for given τ and T we find $\mathcal{N}(b) \sim b^{-1}$ or $d = 1$. *Local:* $bc < \Delta Y$. Hence, as $\Delta Y \sim (b\tau)^{1/2}$ (see [7] above), we find $\mathcal{N}(b;\ c,\ \tau) = (\Delta Y/bc)T/b\tau = (b^{1/2}\tau^{1/2}/bc)T/b\tau \sim b^{-3/2}$ or $d = 3/2$— for given values of c, τ, and T. Summarizing: The local or high-resolution dimension of the profile amounts to $d = 3/2$. The global or low-resolution dimension, $d = 1$, indicates that the global profile is a *non*-fractal curve—it is of no interest to us since "global" demands viewing the profile under "very poor resolution:" The object appears as an ordinary curve.

[9] For a real experiment, see D. Romeu, A. Gómez, J. G. Pérez-Ramirez, R. Silva, O. L. Pérez, A. E. González, and M. José-Yacamán, Surface fractal dimension of small metallic surfaces, *Phys. Rev.* **57,** 2552 (1986).

[10] The rate of change of surface area ε^{2-d} with yardstick size ε, $-(d/d\varepsilon)\varepsilon^{2-d} = -(2-d)\varepsilon^{1-d}$, is zero for a classical ($d = 2$) and ever increasing for the surface fractal ($d > 2$) at ever smaller ε. (To render the differential quotient a positive quantity, a minus sign is conveniently placed in front of it.)

[11] Consider as initiator a large paper tissue on the lab bench. Then perform contraction-shift operations on the paper by crumpling it up (perhaps the experiment failed) ever more tightly. The resulting ball obviously fills volume more extensively than the original, flat tissue. It is also more useful: For instance, it serves as a more effective sponge—mopping up the mess—than its flat initiator and consists indeed of contiguous surface elements of all (linear) size from *small* (depending on the level of frustration) to *large* (diameter of the paper ball itself).

[12] P. Pfeifer and M. Obert, *Fractals: Basic Concepts and Terminology,* in *The Fractal Approach to Heterogeneous Chemistry,* D. Avnir, Ed. (Wiley, Chichester, UK 1989), p. 11.

[13] Some time after this chapter was completed, I fell upon a reference mentioning the surface dimension of sandblasted metals: D. Avnir, D. Farin, and P. Pfeifer, A Discussion of Some Aspects of Surface Fractality and of its Determination, *New J. Chem.* **16,** 439 (1992); see Table 1. The metal surface dimensions after treatment: 2.53–2.67.

2

FRACTAL SURFACE ASPECTS OF ADSORPTION AND PERMEABILITY

2.1 INTRODUCTION AND GENERALITIES

After the preceding general and introductory Chapter 1, we come to the core of the text, discussing applications of fractals to chemistry. We begin with fractal aspects of surfaces and pore structures, phenomena that not only play an important role in the technology and fundamental research of heterogeneous catalysis but also broaden fractal theory introduced in the previous sections.

For catalytic purposes, surface properties are usually classified by physisorption and chemisorption according to weak or strong adsorbate–adsorbent interactions, respectively. It is then convenient to discuss these separately, commencing with physisorption. Here, in order to obtain sufficient activation of a reaction that passes through a physisorption step, it is advantageous to have adsorbents with high-surface area and/or porosity of the order of hundreds of square meters per gram. Cramming such a large area into a mere gram of material is astonishing; I often wondered whether my colleagues in Ford Catalysis Research gave this a second thought. To draw attention to it, I considered stepping up to the podium during one of our weekly seminars, holding up a test tube containing a few cubic centimeters of a high surface area powder (γ-Al_2O_3), scattering its contents haphazardly to the floor and asking the audience: *"Where is* the vast surface"? Of course I never did but the amount of scattered powder would not cover the floor of even a small conference room. Yet, 100 m^2 or so surface area *is* found therein, as known from standard BET (Brunauer–Emmett–Teller) isotherm experiments using the area of the N_2 molecule (16 $Å^2$) as surface yardstick measure. Consequently, the effective surface of the particular adsorbent must be expected to show a rather *irregular* and *convoluted* shape to account for the large surface area within that tiny volume of material.

But how to measure and characterize this in the laboratory? Figure 2.1 demonstrates the concept—a probing or scaling experiment by *different*-sized adsorbate molecules to ascertain the degree *and* the extent of adsorbent surface irregularities. Figure 2.2 shows corresponding results from its application to experimental physisorption data. The procedure seems simple: (a) A charcoal adsorbent is covered to a monolayer with a series of random-coiled styrene–methyl methacrylate polymers of known radius of gyration r_G. (b) Their number $\mathcal{N}(r_G)$ (per gram of adsorbent) is counted by some analytical technique, and (c) plotted versus r_G [1]. Because we are mainly interested in the invariant quantity d—the dimension or space-filling capacity of the adsorbent—we employ a log–log scale of the raw data (●) [see Eq. (1.1)]; as Fig. 2.2 demonstrates, they fall well onto the linear least-squares regression line (■) of coefficient of determination $= 0.951$—indicating that 95.1% of the data points are falling within the result of the regression analysis—and of slope $s = -2.78$. We therefore infer that the *adsorbent is a fractal surface* [2] within linear range $25 \leq r_G \leq 250$ (Å) because, first, its surface dimension $d = 2.78$ is noninteger and, second, (greatly) exceeds that of a classical surface dimension $d = 2$. We notice that the number of adsorbate molecules $\mathcal{N}(r_G)$ of linear size r_G required for monolayer coverage of the adsorbent grows faster with decreasing r_G than the corresponding number for classical surfaces. (The latter situation is indicated in Fig. 2.2 by two bordering lines of slope $s = -2$.)

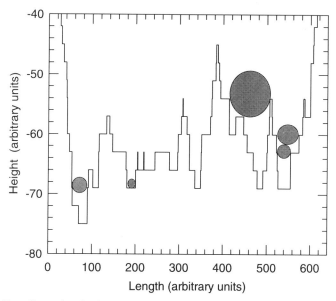

Fig. 2.1 Two-dimensional schematic of measuring (covering) an irregular surface (adsorbent) by different-size adsorbate molecules (hatched circles). The smaller the adsorbate, the larger is the total area it probes on irregularity scales corresponding to its own diameter [see Eq. (1.1)]. Compare with Fig. 1.1(*b*).

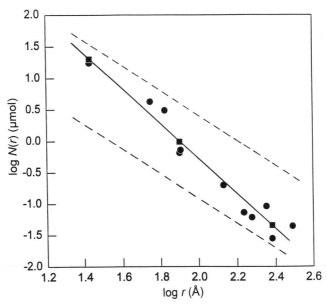

Fig. 2.2 Log–log (decimal) plot of μmol of randomly coiled styrene–methyl methacrylate copolymers (monolayer) versus radius of gyration r_G (Å) on charcoal (●). Molecular weight of the adsorbate ranged from 7400 (r_G = 25 Å) to 4.25 × 10⁵ (r_G = 250 Å). The slope of the linear regression line (■) amounts to −2.78. The two dashed lines represent classical surfaces, slope = −2, drawn at arbitrary positions. [Reprinted with permission from D. Avnir, D. Farin, and P. Pfeifer, *J. Chem. Phys.* **79**, 3566 (1983), Fig. 6, Copyright 1983 American Institute of Physics.]

We have here a clear experimental demonstration that the Brownian-walk–fractal-surface model discussed in Section 1.2 is indeed a realistic, albeit crude approach to describe the real situation [3].

2.2 ANALYSIS OF FRACTAL SURFACES AND PORE STRUCTURES

Before continuing, we ought to stop here for a moment to realize that the aspects of the fractal adsorbent surface tiling method, discussed in Section 2.1, are quite revolutionary. Since long, efforts to analyze adsorbent surfaces have concentrated on a number—the specific surface area by the standard N_2–BET method from adsorption isotherm measurements with adsorbate species N_2. Why then use different-sized adsorbates? Would they not have merely yielded a smaller adsorbent surface or pore area as larger adsorbate molecules do not register smaller adsorbent irregularities?

But, as we have just learned, if surface irregularity prevails at *smaller* and *smaller* scales over some wide range—as certainly implied by the scaling plot of Fig. 2.2—the old classical concepts are no longer valid. The venerable N_2–BET method does not tell the whole story as new effects become apparent: Fractal adsorbents *discriminate* between molecular adsorbate species; from a mixture of different-sized adsorbates, the fractal γ-alumina surface of Fig. 2.2 takes up more of smaller than of larger molecules by a factor $r_G^{-d+2} \sim r_G^{-0.78}$. For example, coverings by molecules with radius of gyration stretching over one decade, say from $r_G = 2$ down to $r_G = 0.2$, engender here a sixfold larger uptake of the smaller species, $(0.2/2)^{-0.78} \approx 6$, than that of a classical adsorbent surface of $d = 2$, $(0.2/2)^{-2+2} = 1$.

Figure 2.3 demonstrates a technical application of the consequences of surface fractality, namely, a fractal filter for removal of atmospheric pollutant NO_x: Note that filter efficiency amounts to 90% for the highest dimensioned filter ($d = 2.9$) compared to only 72% efficiency extrapolated to a filter of classical surface dimension $d = 2$ [4].

However, a successful outcome of surface coverage experiments requires precise extrapolation procedures to monolayer coverage [5]. Furthermore, some inherent ambiguities of this tiling method cannot be overlooked [6], such as (a) lack of precise information on the size of adsorbate molecules and (b) neglect of adsorbent–adsorbate and adsorbate–adsorbate interactions. Therefore, adsorbent tiling techniques were devised that largely avoid some disadvantages of covering an adsorbent by a series of adsorbate molecules of different, and perhaps uncertain molecular size. One such approach, aptly termed inverse (or perhaps "reverse") tiling procedure, ought to be freer of uncertainties yet keeping the transparency of a

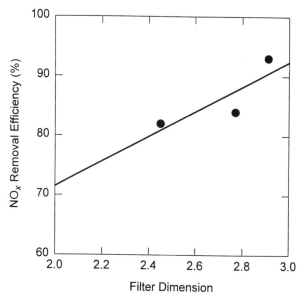

Fig. 2.3 Percent efficiency for NO_X removal from a gas stream by a fractal filter, as function of the filter material dimension. [Reprinted with permission from D. K. Ludlow and T. P. Moberg, *Analyt. Instrum.* **19**, 113 (1990), Fig. 4, courtesy of Marcel Dekker, Inc.]

tiling method. To understand the principle, it is instructive to first recall a basic relation for computing surface area $S(\varepsilon)$: We multiply number–size distribution $\mathcal{N}(\varepsilon) \sim \varepsilon^{-d}$ of surface elements of linear size ε [Eq. (1.1)] by surface element area ε^2,

$$S(\varepsilon) = \eta(\varepsilon)\mathcal{N}(\varepsilon)\varepsilon^2 \sim \varepsilon^{-d+2} \tag{2.1}$$

Equation (2.1) implies that (within some average factor $\eta(\varepsilon)$ [7]) surface area $S(\varepsilon)$ is only independent of diameter or radius ε of the adsorbate molecule if d is classical ($d = 2$). Adsorbent surface area $S(\varepsilon)$ of fractal surfaces ($2 < d < 3$) increases faster under decreasing adsorbate diameter ε than adsorbent surface S of a classical surface. Now, the (statistical) self-similarity or *scale-invariance* of a surface fractal—well demonstrated by the ease to see a "mountain range" in the weathered crystalline outcrop of a rock once absolute scales are hidden (see Figs. 2.4 and 2.5)—implies that it is immaterial whether linear scale parameter ε pertains to adsorbate or to adsorbent. Therefore, instead of using a series of *different-sized adsorbates* to tile the *same adsorbent* to monolayer coverage, we use the inverse scheme and sieve the adsorbent into fractions of *different-sized adsorbent* particles and tile each size fraction of the adsorbent to monolayer coverage with the *same adsorbate* molecule (usually N_2) at a *fixed* volume of adsorbent [8].

Fig. 2.4 Close-up photograph of a free-standing rock (lens: ZUIKO 24 mm, 1:2).

Fig. 2.5 Entrance to C. C. Little Building, University of Michigan Central Campus, Ann Arbor (with permission).

Figure 2.6 illustrates the theoretical basis of this approach on the example of a deterministic, self-similar fractal. Again, in order to keep the graphics simple, respective profiles of surfaces are shown (see Fig. 1.5): The adsorbent is conveniently represented by a segment of the second stage of the so-called Koch curve [9], a deterministic, self-similar fractal of dimension $d = \log 4/\log 3 \approx 1.26$. The monolayer of adsorbate is rendered by thinner lines of dimension $d = 1$; contact points with the Koch curve are accentuated by small square dots.

Upper (*a*) and lower (*b*) portions of Fig. 2.6 are drawn to demonstrate the principle of inverse tiling with constant adsorbate size ε conveniently set to unit length. Two profiles of the surface, of radius $3^{1/2}$ and $3 \times 3^{1/2} = 3^{3/2}$ are shown, scaled by an obvious factor 3. Note that the covering (or measure) of the different-sized adsorbent particles by the same adsorbate is economical: No ambiguities (overlap of adsorbate) are introduced (see *Hausdorff measure* in Glossary 1). As shown, adsorbent–particle coverings are accomplished with two tiles of total length $2\varepsilon = 2$ for the smaller, eight tiles of total length $8\varepsilon = 8$ for the three-times larger adsorbent particle.

We quantify this covering in terms of a mass–distance relation, a fundamental scaling relation between a property of the object—such as mass (M), volume (V), area or surface (S), length (L), density, and so on—and distance R (diameter, radius)

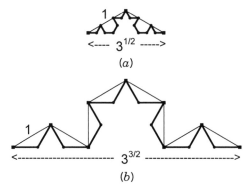

Fig. 2.6 Two-dimensional scheme of the inverse tiling method. The profiles of the adsorbent particle surfaces are represented by the second stage of the Koch curve. The profile of the (single) adsorbate molecule is rendered by thin lines of unit length. (*a*): Adsorbent particle radius $R = 3^{1/2}$. (*b*): Adsorbent particle radius $R = 3 \times 3^{1/2} = 3^{3/2}$.

on the object. Remembering that Eq. (1.1) admits any dimension d in the range $0 \leq d \leq 3$, we give a mass–distance relation for a fractal in the form [10]

$$M(R) \sim R^d \qquad (2.2)$$

Inserting adsorbent diameter $R = 3^{1/2}$ and $R = 3^{3/2}$ into Eq. (2.2), Fig. 2.6 shows that profile length L_C scales with distance R as (recall that we set $\varepsilon = 1$)

$$L_C(R = 3^{1/2}) \sim R^d \sim (3^{1/2})^d = 2 \qquad (2.3a)$$

$$L_C[R = 3(3^{1/2})] \sim (3^{3/2})^d = 8 \qquad (2.3b)$$

respectively.

To emphasize, the step from Eq. (2.3a) to Eq. (2.3b) is simply an application of a change of scale of the independent variable R by multiplicative constant b: $R \Rightarrow bR$. Then, $L_C(R) \sim R^d$ is modified to [11]

$$L_C(bR) \sim (bR)^d \sim b^d R^d \sim b^d L_C(R) \qquad (2.3c)$$

$$L_C(R) \sim b^{-d} L_C(bR)$$

Choice $b = 3$ leads to $L_C(3^{1/2}) = 3^{-d} L_C(3^{3/2})$, or $2 = 8 \times 3^{-d}$. Hence, log $2 = -d(\log 3) + \log 8$, or fractal profile dimension $d = \log 4/\log 3$, identical to that form the standard tiling method [8]—as it must, if we did it right.

Summarizing, the working principles of the inverse tiling method are based on the scale invariance of a self-similar fractal object [Eq. (2.2)]; by now it should be quite obvious that the property of self-similarity is of great utility in describing fractal objects. For our immediate purposes, it leads to a viable alternate adsorbent surface tiling technique permitting us to navigate around some of the questionable assumptions of the standard tiling procedure.

Figure 2.7 illustrates experimental results [12] probing the surface characteristics of periclase. But what do we plot in this reversal method [13]? We plot the area of N_2 adsorbate molecules per unit mass (volume) of the particular adsorbent, $S(R)/R^3 \sim R^d/R^3 \sim R^{d-3}$ [see above and Eq. (2.2)], with quantity R designating the diameters R of its sieved fractions. Note then that we essentially scale a fractal surface of an adsorbent, R^d, with its volume (or mass) R^3. Therefore, slope $s = d - 3 = -1.05 \approx -1$ immediately tells us that the periclase surface is classical: The ratio of surface to volume scales with R as R^{-1}.

So, periclase is then *not* a surface fractal. But we must keep in mind that the statement is not absolute: Fractal aspects do not deal with absolutes. Periclase is not a surface fractal within experimental diameter 3–22 μ (see Fig. 2.7) but it may well be a surface fractal over linear measuring sizes outside this range and not readily accessible to our sieving technique, or available under the employed fragmentation methods generating the various-sized periclase fractions, or just out of scope of practical interests in chemistry.

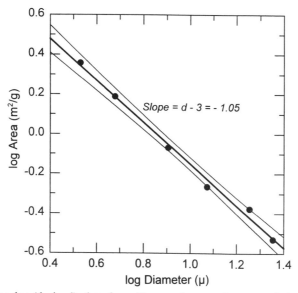

Fig. 2.7 Log–log (decimal) plot of specific monolayer surface area of nitrogen on six size-sieved fractions of periclase as function of their diameter, showing application of the inverse tiling method. The straight line is the least-squares linear regression curve, with its 95% confidence limits, through the data points. Range of periclase fraction size: 3.4–22 μ. [Reprinted with permission, D. Avnir, D. Farin, and P. Pfeifer, *J. Coll. Interface Sci.* **103,** 112 (1985), Fig. 2, Academic Press, Inc.]

2.3 FRACTAL ASPECTS OF PHYSISORPTION PHENOMENA

2.3.1 Gas Adsorption by Microporous Solids

Large specific surface area in many adsorbent materials arises from the interspaces of small, aggregated *impermeable* constituent particles, that is, from the wall area of their *pore volume*. Two limiting situations come to mind: microporosity, that is, pore sizes below approximately 20Å, and zeolite pore structure. Microporosity, in particular, is of widespread occurrence in industrial carbonaceous adsorbent materials such as activated chars. Microporosity distorts dispersion forces between adsorbate and adsorbent into the direction of increased adsorption due to the proximity of opposing surface elements within the narrow pore structure. Thus, fractal theory needs to be tailored to any prevalence of hydrostatic forces because in this situation the number of adsorbed molecules, N, is much greater than monolayer coverage N_m. (Actually, the phenomenon is familiar to anybody who waters a flower pot. Quite a bit of water disappears from sight by moving into the finer and finest interstices between soil particles.)

Appropriate theory [14] relates fraction $\theta = N/N_m$ with adsorbate vapor pressure ratio P/P_0 of equilibrium and saturation pressures, respectively, as

$$\theta = N/N_m \sim [\ln(P_0/P)]^{d-3}   2 < d < 3 \tag{2.4}$$

To appreciate Eq. (2.4), we consider the classical situation of $d = 2$ by (a) taking the adsorbate as a slab of liquid wetting the pore surfaces, (b) assuming that van der Waals dispersion energies determine the difference of the chemical potential between film and bulk liquid, (c) setting the former to be that of its coexisting vapor at pressure P, (d) the latter to be that of its coexistent vapor at saturation pressure P_0. The well-known outcome in terms of a pressure-dependent film thickness $\xi(P)$ reads $\xi(P) \sim [\ln(P_0/P)]^{-1/3}$ or, in terms of a pressure-dependent film volume, $V(P) \sim [\ln(P_0/P)]^{-1}$, respectively—a result predicted by Eq. (2.4) with $d = 2$.

Experimental exploitation of Eq. (2.4) is displayed in Fig. 2.8(a) by N_2 isotherms at 77 K on pitch-based activated carbon fibers [15]. As adsorbent loading is near complete at the very low $P/P_0 \approx 0.001$ (type I isotherm)—attesting to the prevalence of microporosity—we indeed find that the plots of $\ln(\theta)$ versus $\ln[\ln(P_0/P)]$ [Eq. (2.4)] are rather flat over most of the range of useful (very low) P_0/P or (high) P/P_0. To obtain dimension d, the range of fractional filling θ is partitioned into sections of increasingly closer approach toward the limit $P/P_0 = 1$, namely, 0.5–1.0, 0.7–1.0, and 0.8–1.0, and each segment of θ examined for constancy of dimension d within its range. Not surprisingly, the fractal dimension varies from $d = 2.5$ for $\theta \approx 1$ to $d \approx 2.0$ at $\theta = 0.5$, only staying constant at the ranges closest to $\theta = 1$, for instance, $0.8 < \theta < 1$. Figure 2.8(b) shows a representative plot of d within $0.7 < \theta < 1$, demonstrating a reasonably constant dependence of d on θ. Hence, $d = 2.6$ is accepted as the pore dimension of the pitch-based carbon fibers.

How reliable is this result? Note here that the domains $0.7 < \theta < 1$ of apparent

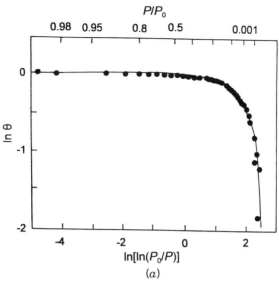

Fig. 2.8(a)   Nitrogen type I isotherm on pitch-based carbon fibers at 77 K plotted bilogarithmically (natural log) according to Eq. (2.4). [Reprinted with permission from K. Kaneko, M. Sato, T. Suzuki, Y. Fujiwara, K. Nishikawa, and M. Jaroniec, *J. Chem. Soc. Faraday Trans.* **87,** 179 (1991), Fig. 4, The Royal Society of Chemistry.]

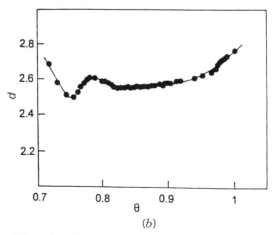

Fig. 2.8(b)   Plot of dimension *d* versus fractional coverage, computed from the data of Fig. 2.8(a). [Reprinted with permission from K. Kaneko, M. Sato, T. Suzuki, Y. Fujiwara, K. Nishikawa, and M. Jaroniec, *J. Chem. Soc. Faraday Trans.* **87,** 179 (1991), Fig. 7c, The Royal Society of Chemistry.]

validity of Eq. (2.4) are rather narrow. Estimating the ratio of upper and lower cutoffs of the valid scaling range by the ratio of film thickness $z_{MAX}/z_{MIN} = \{[\ln(P_0/P_{MIN})]/[\ln(P_0/P_{MAX})]\}^{1/3}$ [14], we get $z_{MAX}/z_{MIN} \approx 3$—even with generous values of $P_{MIN}/P_0 = 0.5$, $P_{MAX}/P_0 = 0.98$ [see Fig. 2.8(a)]. Clearly, more extensive laboratory work is required to refute objections that were raised against the validity of fractal dimensions obtained by this analytical method [16]. Perhaps the effects of multilayer adsorption due to the strong hydrostatic forces in the micropores of the substrate hinder satisfactory application of a surface tiling approach.

Another expression for fractal pore area determination by adsorption isotherms is based on the standard tiling relation [Eq. (2.1)] $S = \mathcal{N}(\varepsilon)\varepsilon^2 \sim \varepsilon^{-d+2}$ but relates linear tile size ε to the radius of curvature of the meniscus ε_M of a convenient condensed-phase adsorbate, usually N_2. Indeed, from the thermodynamic relation between ε_M and vapor pressure ratio P/P_0,

$$\varepsilon_M \sim 1/[T \ln(P_0/P)] \qquad P \leq P_0 \qquad (2.5)$$

we see that radius ε_M increases with increasing pressure P. The advantages of this approach seem obvious because Eq. (2.5) simulates the standard tiling method under a series of different size adsorbates on the basis of a *single* inert molecular species, with its negligible adsorbate–adsorbate and adsorbate–adsorbent interactions [6]. Then, substituting Eq. (2.5) into Eq. (2.1) yields surface area S,

$$S \sim [\ln(P_0/P)]^{d-2} \qquad (2.6)$$

under the assumption that capillary forces in the micropores greatly exceed contributions from adsorption forces. Consequently, the approach is essentially based on the same premises as that of Eq. (2.4). Indeed, forming the classical limit $d = 2$, Eq. (2.4) implies the well-known result that the ratio of area/volume increases with radius R of the object as R^{-1}, whereas Eq. (2.6) implies that classical surface area ($d = 2$) is independent of tiling yardstick size ε [17, 18].

Experimental evaluation of Eq. (2.6) is displayed in Fig. 2.9(a) by the cover N of N_2 at 77 K on activated charcoal (intermediate or transitional pore sizes) [18]. There is a slight hysteresis between adsorption and desorption runs, implying irreversible components within the overall mechanism. Figure 2.9(b) shows the scaling plot of ln (S) versus ln $[\ln(P_0/P)]$ [Eq. (2.6)], yielding slope $s = d - 2 = 0.7$ and thus $d(\text{adsor}) = 2.71$, $d(\text{desor}) = 2.73$, with lower and upper cutoff, $10 < \varepsilon_M < 200$ Å (vertical arrows).

Here, it is certainly remarkable that the range of ε_M of 10–200 Å seems so much wider than that for probing the pore surface of carbon fibers [Fig. 2.8(b)], which stretched over a mere factor of about 3. This question is particularly opportune as the basis of the two analytical methods is essentially the same, as mentioned above. Perhaps a "devil's advocate" will argue—and not entirely unjustified—that the apparently wide linear range of 10–200 Å in the scaling plot of Fig. 2.9(b) is illusory, maintaining that there is no *strictly linear* region as the data seemingly follow a smooth course with no discernible crossover (points of inflection).

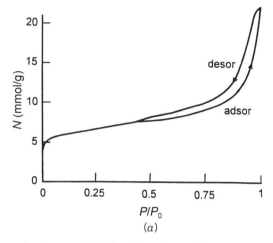

Fig. 2.9(a) Nitrogen isotherms at 77 K for adsorption and desorption on activated charcoal. [Reprinted with permission from A. V. Neimark, *JETP Lett.* **51,** 607 (1990), Fig. 1(a), Copyright 1990 American Institute of Physics.]

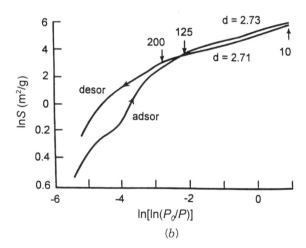

Fig. 2.9(b) Log–log (natural) plots of specific surface area versus the logarithm of the inverse relative adsorbate pressure for adsorption and desorption, according to Eqs. (2.5) and (2.6). Fractal dimensions for adsorption and desorption are attached to the curves and their linear range of validity pointed by the arrows, with numerical values (in Å) indicated. [Reprinted with permission from A. V. Neimark, *JETP Lett.* **51,** 607 (1990), Fig. 1(b), Copyright 1990 American Institute of Physics.]

We conclude: Credibility and impact of the results of special fractal surface tiling methods that probe adsorption in the presence of strong capillary forces would be greatly enhanced if accompanied by an objective and thorough error analysis. At present reading, such a procedure is rarely followed.

2.3.2   Gas Adsorption by Zeolites

Zeolites, widely applied in chemical research and its industrial applications, are taken to be the compounds of prima facie regular pore structures. However, it appears that surface analysis, using the various tiling methods already discussed in this chapter, frequently implies considerable fractality of zeolite surfaces upon rather mild treatment—not a characteristic we would have expected. Indeed, published results yielding surface dimensions $d \approx 2$ of treated zeolites are relatively rare. The cause leading to fractal character—it turns out to be treatment-induced dealumination—is evidently important to know about and to quantify.

For instance, measurements of adsorption isotherms on zeolite Na-Y (Si/A1 = 2.47), using the standard tiling method with alcohol adsorbates methanol, ethanol, 2-propanol, and 2-methyl-1-propanol, yielded typical type I isotherms implying capillarity [19]. The resulting dimension d, obtained from the slope of plots of log (number of alcohol molecules adsorbed) versus log (linear size of adsorbate), indicate a classical surface with $d = 2.03 \pm 0.05$—an expected outcome. However, transforming the zeolite by ion exchange into a 92% [NH_4, Na-Y] daughter (Si/A1 = 2.45) apparently induces fractality: Surface analysis now indicates $d = 2.43 \pm 0.04$, a considerable excess over its parent compound ($d = 2.03$). Note that the original Si/A1 ratio has barely changed upon treatment; hence, the effect would *not* be picked up with sufficient discrimination by elemental analysis. Similarly, ^{29}Si nuclear magnetic resonance (NMR), X-ray diffraction (XRD) as well as Fourier transform infrared spectroscopy (FTIR) observations of several vibrational transitions in the 460–1140-cm^{-1} range in the treated material do not indicate anything unusual [19].

It therefore appears that zeolite treatment (even under the mildest conditions) removes A1, generating framework vacancies and nonframework interstitial A1 species. Interestingly, portions of the changes are reversible: Realumination of the [NH_4, Na-Y] daughter (by aqueous KOH) returns dimension $d = 2.03$ of the parent compound. This outcome implies that the surface fractality of the modified structure arises from the *extra-frame* aluminum sites [20].

Reports on the other parent-modified daughter zeolites invariably disclose considerable structural surface fractality [21] on the basis of tiling analysis. Consequently, the global topology of the symmetrically shaped, long-range order zeolite structure may be superposed by sites of *local* fractal geometry. Practical implications of these findings can hardly be overlooked. Foremost, it should not be assumed a priori that a zeolite structure is solely Euclidian on the merits of its regularity and symmetry; very mild treatment apparently introduces local, extra-frame fractal sites. At the very least, fractal analysis of isotherm data alerts to such treatment-induced structural changes that, apparently, are not readily registered by the usual analytical methods.

In concluding this section, applications of pore *volume* measurements to characterize pore volume distributions of porous adsorbents such as zeolite, coal, and so on, are discussed. Experimentally, methods of computer-controlled high-pressure mercury intrusion have been used [22] to determine porosity parameters such as pore radii, cumulative pore volume, pore surface area, and differential pore volume for the pore size range 75–75000 Å. (Pore volume distributions for the range 12–450 Å are obtained by any of the N_2 tiling methods as discussed above.) Fractal aspects of distributions of pore radii (ρ) are obtained by scaling the differential pore volume $-dV/d\rho$—an area [8, 23]—with pore radius ρ:

$$S = -dV/d\rho \sim \rho^{2-d} \qquad \rho \Rightarrow 0 \qquad (2.7)$$

Note that this relation is simply the number–size scaling relation Eq. (1.1) multiplied by the pore area element ρ^2 {see Eq. (2.1) and [12]}. For nonwetting intrusion methods (e.g., by Hg), area S is related to the differential volume change with external pressure P by [24]

$$S = dV/dP \sim P^{d-4} \qquad (2.8)$$

Figure 2.10 shows an application of Eqs. (2.7) and (2.8) to Hg-pore intrusion data of a fine grain pentasil-type zeolite [25]. Two apparent pore size ranges of different

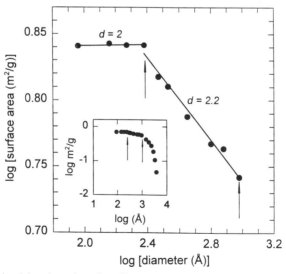

Fig. 2.10 Decimal log–log plot of scaling pore surface area with pore radius from Hg-intrusion measurements of a fine grain pentasil-type zeolite. Two different ranges of fractal behavior, their cross-over points, and respective fractal dimensions are indicated (least-squares linear regressions). The inset shows the entire data range (units of 10 m²/g). [Data from H. Spindler and M. Kraft, *Catal. Today* **3**, 395 (1988).]

fractal dimension $d = 2 - s$ are noticeable from slopes s of the log $[S(\rho)]$–log(ρ) plot, namely, $d = 2$ over range $75 < \rho < 250$ Å and $d = 2.2$ over $250 < \rho < 1000$ Å. At pore sizes larger than 1000 Å, the slope drops precipitously to large negative values, as shown by the inset of Fig. 2.10.

However, the fractal dimension $d = 2.2$ (and higher) at pore sizes exceeding 250 Å is probably not real. Recall that Eq. (2.7) requires the limit toward smallest pore radii [8]: It can be argued that, first, pore radii within a range of 250–1000 Å do not meet this criterion, second, that the support of numerical value $d = 2.2$ is too narrow (factor of 4). Hence, it appears that dimension $d = 2$ for the smallest pores of the zeolite is the only acceptable result of this study. Certainly, it would be very useful to find out why the Hg-intrusion method leads here to such unreasonably high dimensions ($d \approx 3$) at largest pore radii [26].

2.3.3 Adsorbency of Porous Silicas

The surface characteristics of adsorbent materials made of silica have attracted considerable attention, last but not least because of wide industrial use of silica-based formulations in manufacturing, construction, and packaging. Silica compounds exhibit a wide range of surface dimensions that depend critically on composition, preparation, and treatment.

In the following sections, we will discuss surface probing of silica and, in addition, include a demonstration on the effects of preloaded films of small molecules on fractal surface characteristics of the silica substrate [27, 28]. Therefore, we now determine the surface fractality of silica with probe molecule (N_2) looking, so to speak, at the underlying substrate *through* the layers of preloaded film material.

First, some definitions of appropriate parameters and scaling relations. Denoting thickness of the preloaded film by z, its area by $A(z)$, and volume by $V(z)$, respectively, yields the elementary relations

$$A(z)dz = dV(z) \tag{2.9a}$$

$$A(z) = dV(z)/dz \tag{2.9b}$$

where dV, dz are small increments of film volume V and film thickness z. (It is assumed that the outer surface of the preloaded film is point-by-point equidistant from the adsorbent surface, at least on average.) For an adsorbent surface of dimension $2 < d < 3$, recall that the number $\mathcal{N}(z)$ of self-similar pieces of linear size z making up the whole is given by $\mathcal{N}(z) \sim z^{-d}$ [Eq. (1.1)]. Hence, film volume $V(z)$ is given by

$$V(z) \sim \mathcal{N}(z)z^3 \sim z^{3-d} \tag{2.10}$$

Subsequently, its area $A(z)$ is obtained from Eq. (2.10) by differentiation [Eq. (2.9b)]:

$$A(z) \sim z^{2-d} \tag{2.11}$$

[note that Eq. (2.11) resembles Eq. (2.7)]. Solving Eqs. (2.10) and (2.11) for z gives film thickness $z = A(z)^{1/(2-d)} = V(z)^{1/(3-d)}$. Then, film area $A(z)$ is expressible solely by film volume $V(z)$ and surface dimension d,

$$A(z) \sim [V(z)]^{(2-d)/(3-d)} \tag{2.12}$$

What has been accomplished? First, we need to be clear to which object an experimental surface dimension d of the nitrogen covering data refers. Clearly, it refers directly to the surface dimension of the preloaded film and only indirectly to that of the naked adsorbent: Only in the limiting situation of a classical adsorbent surface are surface dimensions of film deposited and of adsorbent covered identically $d = 2$. Second, we expect the analytical plots of log $[A(z)]$ versus log $[V(z)]$ to be linear and of slope $s = (d-2)/(d-3)$. Third, as $2 < d < 3$, we see that $-\infty < s < 0$—the upper value implying $d = 2$. Thus, a bonus of this "sandwich" approach entails from a large sensitivity of dimension d to small changes in experimental slope s.

Figure 2.11 demonstrates the capability of the method by isotherms of N_2 adsorption (77 K) on preloaded films of water and n-heptane, respectively, either ad-

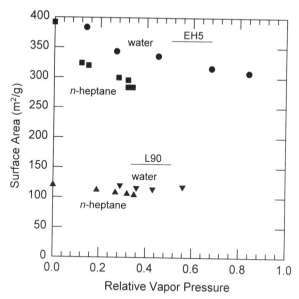

Fig. 2.11   Surface area of water and n-heptane films preloaded on fumed silica Cab-O-Sil EH5 (rough grade) and Cab-O-Sil L90 (smooth grade), as probed by N_2 adsorption at 77 K. Data points: water on EH5 (●) and on L90 (▼), n-heptane on EH5 (■) and on L90 (▲). Vapor pressures (relative to saturation) pertain to the loadings of the respective film material. [Reprinted with permission from P. Pfeifer, G. P. Johnston, R. Deshpande, D. M. Smith, and A. J. Hurd, *Langmuir* **7**, 2833 (1991), Fig. 3, Copyright 1991 American Chemical Society.]

sorbed at ratio P/P_0 [29] on two fumed silica samples (Cab-O-Sil). Results vary with respect to combinations of the molecular nature of preloaded film and the roughness grades, L90 (smoother) and EH5 (rougher), of the silica adsorbent. First, note that on the rougher grade (EH5) the N_2-probed effective area is smaller for preadsorbed n-heptane than for preadsorbed water. The explanation is evident: n-heptane, being a larger molecule than H_2O, hides the smaller pores of the silica adsorbent, thereby rendering them invisible to the N_2 probe. The effect is, understandably, less pronounced on the smoother grade (L90) of the silica adsorbent. Second, under increasing P/P_0 [29] the amount of N_2-uptake decreases faster on the rougher (EH5) than on the smoother (L90) sample—regardless of the nature of the film material (cf. upper and lower portions in Fig. 2.11). How do we account for this? Just picture a snowfall over a rocky terrain surrounding a frozen lake: The rugged parts of the earthen ground are increasingly blanked by the falling snow, first rapidly (its many smaller irregularities), then slower (the fewer, larger ones). On the other hand, the frozen lake surface stays near flat as ever.

Now we consider the experimental d values. Recall, these are the dimensions of the preloaded films, evaluation by Eq. (2.12) yielding $d(\text{L90}) = 1.98$ and $d(\text{EH5}) = 2.10$ within experimental error $\pm$ 0.10 [28, 30]. Then, there is little ambiguity: Closeness to $d = 2$ for preloaded films of two widely different-size molecular film species predicts that the underlying silica adsorbents are not surface fractals. (In sections 5.1.2.1 and 5.1.2.2, this somewhat unexpected outcome will appear in other high-surface silicas, probed by other methods.) We conclude: The surface irregularities are not of a *fractal* nature.

This finding brings up the important point: What is the predicted range of the experimentally ascertainable adsorbent surface irregularity covered by this result? It is estimated from thickness z of the preadsorbed film [30] and yields in these experiments an upper limit of $z \approx 10$ Å, implying that the film consists of several layers of water molecules. The range could, in principle, be extended by higher film loadings; however, this requires larger P/P_0 and may run into capillary effects, obviating a meaningful measure of film thickness z (see Section 2.3.1). Indeed a related, certainly unintended effect of this genre (viz., pore blocking) will occupy us in Section 2.4.3.

Additional and wide-range applications of this method can be envisaged, such as designing desirable *effective* fractal adsorbent surface dimensions by controlled smoothing from preloaded films of appropriate materials, not necessarily volatile.

2.3.4   Generation of Pore Structure: Dehydration of Gibbsite

The processes of particle aggregation leading to fractal pore surface by generation of fractal interior interfaces (few large, more smaller, many smallest) between the touching, impenetrable particles making up a global aggregate have not yet been discussed. Its aspects well deserve their own treatment and will be introduced in Section 3.1.2 further below.

On the other hand, pore structure may be generated in essentially nonporous crystalline materials by forceful, temperature-shocked ejection of lattice-bound molecules: Industrially useful materials, for instance, gibbsite (lattice-bound water) and chalk (CO_2), come to mind. In brief, the technology involves rapid transport of the powdered material through a furnace at preset temperatures and residence times. Subsequently, the product is analyzed for any treatment-induced variations of its surface properties, in particular pore size distributions and pore surface dimensions from the usual log $(-dV/d\rho)$ versus $\log(\rho)$ plots [Eq. (2.7)].

Table 2.1 summarizes recently published results on gibbsite [31] by (a) treatment temperature in centrigrade, (b) analytical composition of the resulting activated alumina sample by mol ratio H_2O/Al_2O_3, (c) percent porosity, (d) BET surface area, and (e) pore dimension, obtained from pore size distribution $-dV(\rho)/d\rho$ calculated by standard methods from the desorption nitrogen isotherms and, subsequently, scaled with pore radius ρ according to Eq. (2.7). It is obvious that the higher the temperature of the heat treatment step, the larger the loss of lattice-bound water, and the closer the dimension of the activated Al_2O_3 to embedding space dimension $d = 3$. Hence, manufacture of activated refractory pore structure materials with desired characteristics of surface fractality are accessible by varying furnace temperature and material residence time.

In general, knowledge of the fractal dimension of an adsorbent predicts its differential capacity of taking up molecular adsorbates of varying size. Such modeling may serve well in the manufacture of adsorbents of selected adsorptivity for certain molecular species (see also Section 2.2). As an example, Fig. 2.12 demonstrates this for the activated Al_2O_3 samples generated from temperature-treated gibbsite (Table 2.1). The procedure is as follows: We know fractal dimension d of the activated alumina at the various treatment temperatures of the gibbsite and its surface area for adsorbate N_2 from the BET measurements. Choosing convenient, nonreactive adsorbates CH_4, n-C_4H_{10}, and n-$C_{32}H_{66}$—their molecular cross sections ε^2 entered on the abscissa of Fig. 2.12—we calculate their monolayer surface area relative to the absolute BET value of the N_2 coverage by simply modifying Eq. (2.1) to $S(\varepsilon^2) \sim (\varepsilon^2)^{-d/2}\varepsilon^2 \sim (\varepsilon^2)^{-(d/2)+1}$. Note that adsorbate methane leads to a surface area of 160 m²/g on the 580 °C activated alumina ($d = 2.27$) and of 240 m²/g on 650 °C activated alumina ($d = 2.40$); for adsorbate n-$C_{32}H_{66}$ these values are reduced to 120 and 150 m²/g, respectively. Tiled surface area on nontreated gibbsite ($d \approx 2$) is seen to be one order of magnitude smaller. Clearly, then, effects of increasing surface dimension d above classical value $d = 2$ are not linear with d: There will be a reminder of this below.

TABLE 2.1 Characteristics of Untreated and Temperature-Shocked Gibbsite [31]

Treatment Temperature (°C)	Mol Ratio H_2O/Al_2O_3	Percent Porosity	Surface Area (BET) (m²/g)	Fractal Pore Dimension
Not treated	3	16	19	2.01
580	1.42	37	173	2.27
650	0.85	51	251	2.40

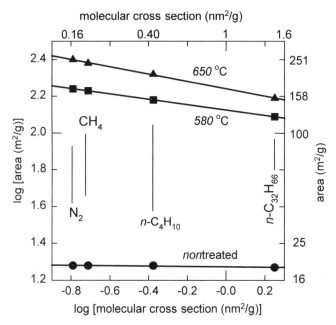

Fig. 2.12 Modeling of the effective surface area of an activated gibbsite by four species of adsorbate molecules of known cross section, demonstrating preferential adsorption of the smaller adsorbate species. The ordinate indicates the (decimal) logarithm of covered specific adsorbent area, the abscissa the logarithm of adsorbate cross section of the four adsorbate species. Linear least-squares slopes s are $s = 0.01$ (nontreated), $s = -0.145$ (580 °C), and $s = -0.200$ (650 °C), with coefficients of determination 0.856, 0.999, and 0.999, respectively. The gibbsite dimensions are listed in Table 2.1. [Dimensional and cross-sectional data from N. N. Jovanović, V. R. Nikolić, T. Novaković, and A. Terlecki-Baričević, *J. Serb. Chem. Soc.* **58**, 229 (1993).]

The aspect of scaling plots such as Fig. 2.12 should now be familiar: On a fractal adsorbent surface, the smaller the probe molecule, the larger the measured area. The smaller species fills more nooks and crannies. On a classical surface, such size dependence does not exist or, at best, stretches over rather limited ranges of molecular adsorbent pore distributions [7]. Considering the nonpolar character of N_2 and the three *n*-alkane molecular adsorbates, it is unlikely that the adsorption process reflects anything but physisorption—reversible fixation of adsorbate under minimal specific adsorbate–adsorbent interactions. We are dealing then with a size effect that depends essentially on the nature of the adsorbent pore size distribution and not on its chemical properties. The latter, prevalent in chemisorption and obviously introducing complications into theory and evaluation of experimental data, will occupy us in Section 2.4.

Fractal analysis is silent on the *structure* of the pores: It only tells us about the pore size *distribution*. Yet, this is the information of most actual value and usefulness for amorphous materials: Only for precisely structured objects (zeolites) would knowledge of their pore structure be of utility.

2.3.5   Adsorptivity of Polymers on Porous Fractals

Adsorption effects of polymers and other chain molecules on surfaces or within pore structures of solid adsorbents play a significant role in diverse systems and processes of chemical interest, such as oil recovery, chromatography, and manufacture of many products of daily use and consumption (food stuff, drugs, packaging materials, coatings, and so on). In addition, the subject has considerable scientific interest as isolated, solvated (swollen) polymer chains assume certain statistical fractal configurations, which may be altered upon adsorption on a solid [32]: Hence, the nature of the adsorption phenomenon on nonclassical surfaces is one of "fractal–fractal" interaction—probing a *fractal by a fractal*. Last but not least, as flexible macromolecules are the only available adsorbates for tiling the larger surface elements of a fractal substrate (see Fig. 2.2), their expected adsorbed-state configurations determines the size of their surface-probing effect.

Figure 2.13 gives a planar schematic of two limiting adsorbed configurations of a linear polymer chain [33] on a portion of a deterministic fractal adsorbent of

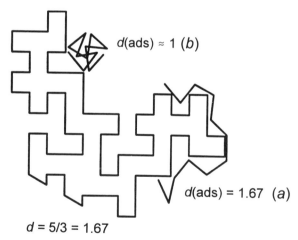

Fig. 2.13 Planar schematic of two limiting adsorbed configurations of a linear chain of degree of polymerization $N = 11$ on the contour of a deterministic mass/pore/surface fractal of self-similar perimeter dimension $d = \log 32/\log 8 = \frac{5}{3}$, a quadratic Koch island. (*a*) $d(\text{ads}) = 1.67$: The solvated chain has unfolded on the adsorbent of $d = 1.67$, probing thereby its fractal perimeter dimension. (*b*) $d(\text{ads}) \approx 1$: The adsorbed chain retains its solvated random-coil configuration at a radius of gyration comparable to the smallest irregularities of the adsorbent profile, thereby no longer probing its fractal dimension but the topological dimension ($d = 1$) of its constituent pieces. Note that the sketch is in terms of perimeter quantities of relevant adsorbate and adsorbent surfaces.

dimension $\frac{5}{3}$ (quadratic Koch island, see pp. 54–55 of [2], Section 1.1). Certainly, the scheme does not exhibit a realistic adsorbent (again we are showing profiles or contours of surfaces) but serves to clarify the defining parameters. First, note that the coiled chains probe adsorbent surface irregularities on a smaller scale than their unfolded chain configurations: The former are characterized by their radius (or diameter) of gyration r_G, the latter by their on-surface diameters 2ς [34, 35], as shown in the figure.

Adsorbate–adsorbent configurations by inherently flexible adsorbates render surface analysis of actual fractal adsorbents more challenging than under rigid adsorbate conditions. Yet, the required modifications of basic theoretical-fractal principles are straightforward: As we now deal with two different adsorbate dimensions, we assign (a) a fractal dimension, $d(\text{sol})$, to the configurational state of the swollen chain (random coil), (b) a dimension, $d(\text{ads})$, to its adsorbed-state chain configuration, and (c) a dimension, d, to the surface or pore structure of the adsorbent proper. Then, we write (inverted) mass–distance scaling relations [see Eq. (2.2)]. First, we relate quantity r_G (radius of gyration of the solvent-swollen chain) and degree of polymerization N through dimension $d(\text{sol})$,

$$r_G \sim N^{1/d(\text{sol})} \tag{2.13}$$

degree of polymerization N being more general than mass M. Second, we relate ς (on-surface radius) and N through dimension $d(\text{ads})$,

$$\varsigma \sim N^{1/d(\text{ads})} \tag{2.14}$$

Now, we eliminate the usually unknown on-surface chain radius ς by (a) modifying Eq. (1.1), $\mathcal{N}(\varepsilon) \sim \varepsilon^{-d}$, to read $\mathcal{N}(\varsigma) \sim \varsigma^{-d}$, (b) solving for ς, and (c) equating it with ς of Eq. (2.14). Hence,

$$\mathcal{N}(\varsigma) \sim N^{-d/d(\text{ads})} \tag{2.15}$$

Equation (2.15) then denotes the number $\mathcal{N}$ of chains of degree of polymerization N, of on-surface length 2ς, and of solvated radius of gyration r_G that are needed to tile the surface of the adsorbent by a monolayer.

We have the choice of replacing N by r_G [using Eq. (2.13)], which gives a relation between adsorbent tiling number $\mathcal{N}$, random-coil radius of gyration r_G (solvated chain), and the three dimensions $d(\text{sol})$, $d(\text{ads})$, and d [33]:

$$\mathcal{N}(r_G) \sim r_G^{-d[d(\text{sol})/d(\text{ads})]} \tag{2.16}$$

Note that the simple parent relation [Eq. (1.1)] is recovered if $d(\text{sol}) = d(\text{ads})$. Let us look at two limiting situations: (a) The coiled polymer chains have *unfolded* upon adsorption: $d(\text{sol}) \Rightarrow d(\text{ads}) = d$. The uncoiled chains probe the surface irregularities at a scale bounded from below by the length of a monomer unit, from above by the on-surface length 2ς [34]. (b) The adsorbed polymer chains *retain* their solvated state due to steric hindrance within the pore spaces of the adsorbent: $d(\text{ads}) = d(\text{sol}) \neq d$. Adsorbed chains are in a nonequilibrium state [36] and the coiled chains probe surface irregularities of the polymer substrate at a scale corresponding to their particular solvated radius of gyration r_G. Figure 2.13 shows this schematically, in terms of perimeter dimensions, for situations (a) and (b).

The approach just described can be extended to the perspective of *adsorbed state* configurations rather than *adsorbent surface* geometry, with the help of Eq. (2.15); this aspect is its greatest promise, interest, and merit. However, we would require information about adsorbent surface dimension d, perhaps from other tiling methods described in these sections. In the absence of such information, it has been suggested to extend general mass–distance scaling relation $M(R) \sim R^d$ [Eq. (2.2)] to read [37]

$$\text{property} \sim \text{scale}^{\beta} \tag{2.17}$$

replacing (a) mass by more general system quantity "property," (b) radius R by "scale," (c) and fractal dimension d by an exponent β. Of course, we then have to investigate, case-by-case, whether exponent β represents a fractal dimension, or a ratio of fractal dimensions as in Eq. (2.16), or implies a more general scaling property. Later, we shall find that the latter is the case in many experimental situations involving solids with fractal distribution of active sites.

2.3.6 A Summary

At the Physical Chemistry/Chemical Physics sessions during the 1987 March meeting of the American Physical Society in New York City, not long after Mandelbrot's work had reached a wider audience, there were sessions that left you breathless from the huge variety of subjects on fractals. Indeed, exuberance sort of ran away from sober reflection: Perhaps exponent β of Eq. (2.17) above was taken too frequently for a fractal dimension without deeper investigation. In retrospect, the new concept filled an all to human desire to reduce complex situations to one simple explanation. Fractals fit the bill perfectly. However, with time, expectations have been trimmed. Fractals are neither simple, *au contraire,* nor do they answer well all the questions initially thought to be accessible through their formalism. On the other hand, realistic applications of fractals to topics where they "deal from a strong hand," such as the large variety of physisorption–interface phenomena already discussed, bring new and unique insight and promise novel applications or have already done so. A cursory sweep of the literature yields a vast surplus of materials just asking for further experimentation to explore hints of surface fractality.

However, the reader familiar with aspects of physisorption phenomena should have noticed that agreement between theory and experiment is far from perfect and, in some instances, not too convincing. It is advisable to chose among the available surface probing methods the one best adapted to the system examined and the one most likely to minimize perturbing effects, such as specific adsorbate–adsorbate or adsorbate–adsorbent interaction. Further, an error estimate ought to be included, particularly if salient points of data evaluations and their predictions rest strongly on values from least-squares slopes that are computed from log–log plots. Statistical software programs routinely compute, at least, the standard deviation or errors of averages and the reader ought to be rightfully weary accepting conclusions or predictions based on small differences of fractal surface dimensions without being able to judge the range of uncertainty of the experimental numbers. It is therefore paramount to strive for boundary conditions allowing the widest range of linear probe scales. If not, it may be that supposedly fractal aspects of physisorption are manifestations of multiple Gaussian-distributed Euclidian pore or surface elements under particular large variances of their size distribution.

I do not share the conviction that a particular analytical approach is preferable [38]: *All* methods have their drawbacks and fractal science is not different here from other undertakings. Finally, if images of the actual pore or surface regions can be taken, the numerical results of fractal analysis will be even more credible, as a glance at some pictures of pore fractals [39] demonstrates.

2.4 FRACTAL ASPECTS OF CHEMISORPTION PHENOMENA

The acceleration of an elementary chemical rate from chemisorption states of the reactants is among the most important steps in heterogeneous catalysis; furthermore, specific catalytic site activities can be used to direct a reaction toward a desired product formation. Just consider the automotive three-way converter on everybody's car, a technological accomplishment juggling steps of oxidation (CO, hydrocarbons to H_2O and CO_2) and reduction (oxides of nitrogen to N_2) in the presence and with participation of various side products [40]. It is no surprise that chemisorption phenomena of the various reactants, intermediates, and products play an important role in the overall functioning of the device [41].

But what are the *fractal* aspects of chemisorption? In fact, they have been around, had been frequently used in catalytic theory and praxis—except that nobody called them by this name. Recall, for instance, "island formation" and poisoning of and by chemisorbed species. Among the best known examples is the Pt-catalyzed reaction of CO and O_2, where (a) site requirements for one adsorbate ($O_2 \Rightarrow 2O$) are affected by site blocking from the other (CO), where (b) product formation (desorption of CO_2) regenerates chemisorption sites, and where (c) any other strongly temperature-dependent reaction steps, such as surface diffusion forming activated complexes, may become rate determining. Hence, even on the smoothest substrate and under uniform vapor-deposition rates of reactants, the distribution of the chemisorption sites may become highly irregular on many length scales. One of the earliest surface computer algorithms, modeling reaction $O + CO \Rightarrow CO_2$ on a regular two-dimensional surface, indeed identified islands of O and CO having perimeters of fractal geometry [42].

Furthermore, chemisorption sites are usually found to lead to appreciably less than monolayer coverage of the chemisorbate: This finding indicates that chemisorption site-occupation distributions on the catalyst may be largely noncontiguous or fragmented. Hence, we a priori expect to find fractal aspects in chemisorption phenomena.

2.4.1 Fractal Scaling Laws for Chemisorption

From the onset, it appears that experimental evaluation and theoretical interpretation of fractal aspects of chemisorption are considerably more involved than physisorption. By the very nature of chemisorption, specific adsorbent–adsorbate interactions are strong. No longer can any nonclassical adsorbent geometry be the *sole* criterion of its degree of fractality; site *geometry* and site *reactivity* are now strongly or even inseparably correlated. Furthermore, whereas fractality of an adsorbent acting by physisorption might be independent of the nature of the adsorbate other than its size (see Sections 2.1 and 2.2), this is no longer the case for chemisorption: Any observed chemisorption fractality of a particular adsorbent is, generally, *adsorbate specific* [43, 44]. Furthermore, most interest and applications pertain to catalytic processes employing an active agent (noble metal perhaps) dispersed on an inert support, the latter frequently refractory, amorphous, and of high specific surface or pore area—hence frequently a fractal (see, e.g., Fig. 2.2).

A little reflection will convince you that this very situation contains aspects that are similar to those discussed above for "sandwich" adsorbents (Section 2.3.3); that is, the technique of preloading of adsorbent by films of some molecular species onto which, subsequently, a surface-probe molecule is deposited: For chemisorption, we merely replace the preloaded film by the dispersed chemisorption agent that subsequently, on its active sites, takes up chemisorbates.

Another instructive comparison comes to mind, namely, that concerning the technique of loading of adsorbent by solvated polymer chains (Section 2.3.5). We found earlier that three dimensions need consideration, (a) dimension d of the naked adsorbent surface, (b) dimension $d(\text{sol})$ of the solvent-swollen polymer chain, and (c) dimension $d(\text{ads})$ of the adsorbed polymer chain. Well, putting an active chemisorption agent on its support often involves deposition from a solution phase, say aqueous Pt nitrate. Clearly, the random distribution of Pt ions in solution is not kept after its deposition on the support: Charge interactions with ionic sites or dangling groups (OH) on the support, physical penetration into a complex pore structure of the support material, and sintering phenomena upon the final drying and calcination steps of the catalyst, and so on, certainly drive the dispersed agent into some rather complex final distribution that differs significantly from its homogeneous solution phase and that of the surface and pore elements of its support material. It stands to reason that the state of the product may not be so readily reproducible, not because the experimenter's work lacked precision but because the preparatory conditions could not be controlled (sensitivity to initial conditions). Whether this has been actively looked at in laboratory work is not known to me, but many instances are recalled where reported chemisorption results, carried out among reputable laboratories under identical-precise conditions, nevertheless turned out to differ more than one would have liked or expected. Consequently, any fractal-theoretical description of chemisorption phenomena can be expected to be difficult, complex, and somewhat ambiguous—at least in most situations.

To approach this heuristically–empirically, for instance, in order to combine the effects of site geometry and site activity but still keeping the simple form of mass (volume/area)–distance scaling relations such as Eqs. (2.2) and (2.10), the specific concept of mass is reinterpreted in terms of the more general "chemisorption capacity," written m_C. Its associated dimension is now called "chemisorption dimension" and given the symbol D_C [45]. Capacity m_C (the property) is conveniently normalized to unit weight or volume of adsorbent; R is again a length or distance (the scale) on the object. Therefore, exponent β is replaced by $\beta - 3 = D_C - 3$ and Eq. (2.17) now reads

$$m_C \sim R^{D_C - 3} \tag{2.18}$$

Keep in mind, however, that Eq. (2.18) and related relations do not necessarily engender that power law exponent D_C is a fractal dimension or, for that matter, a dimension at all [46] (see also Section 2.3.5). This fact underlines the already established aspect that a straightforward fractal interpretation of fractal chemisorption parameters, such as m_C and D_C, is no longer as simple as the evaluation of corresponding quantities in physisorption (see Sections 1.1–2.3).

In the following sections 2.4.2 and 2.4.3, we discuss applications of Eq. (2.18) and of related expressions to chemisorption data from the literature. The results will indicate the usefulness of this scaling approach in spite of inherent ambiguities.

2.4.2   Chemisorption of H$_2$ and CO on Metal Clusters

Molecular hydrogen and carbon monoxide are frequently used to characterize chemisorption phenomena of metals. In particular, the chemisorption data are expected to shed light on the shape and distribution of the metal chemisorption sites. To optimize a heterogeneous catalytic process, one would certainly strive to maximize the number of sites that are active (open) under minimal analytical metal content. Studies of fractal chemisorption behavior of catalysts composed of noble metals dispersed on an amorphous support has therefore been pursued vigorously, results frequently indicating chemisorption dimension range $1.7 < D_C < 2.3$ [37]. Note its inferiority to that found for physisorption, where fractal adsorbent dimensions usually exceed classical surface value $d = 2$ (see Sections 1.1–2.3).

An example for D_C at the lower limit of this range is shown in Fig. 2.14, a scaling plot of H$_2$ chemisorption on Pt dispersed on silica in the form of crystallites of known diameter (from electron microscopy) [47]. Chemisorption dimension $D_C = 3 - 1.33 = 1.67$, falling below classical surface dimension $d = 2$, may signify that the H$_2$ adsorbate sits on crystal sites distributed as *fractal curve* $(1 < D_C < 2)$ as sketched in Fig. 2.15. On the other hand, $D_C = 1.67$ may also arise from a mixture of active sites on (classical) faces of the crystallites $(d = 2)$, its edges $(d = 1)$, or corners $(d = 0)$ [37]. Clearly, the latter outcome has nothing to do with fractal aspects of chemisorption; chemisorption dimension D_C here merely reflects a linear combination of the effects of different, active Euclidian sites on the catalyst.

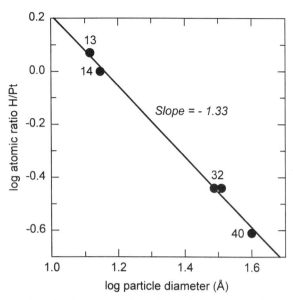

Fig. 2.14 Scaling plot of hydrogen chemisorption on Pt crystallites dispersed on silica. The straight line represents the least-squares linear regression (coefficient of determination = 0.995) through the data points. Slope $s = -1.33 = D_C - 3$ yields $D_C = 1.67$. Numbers at the data points indicate Pt-particle diameter in angstrom (Å). [Reprinted with permission from P. Pfeifer, *Chimia* **39,** 120 (1985), Fig. 6, *Chimia,* Schweizer Chemiker Verband.]

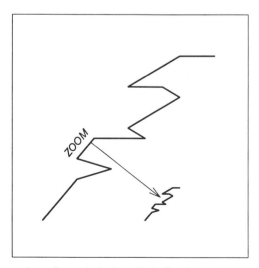

Fig. 2.15 Bird's eye view of a statistically self-similar fractal ridge structure representing chemisorption sites under chemisorption dimension $1 < D_C < 2$ on a facet of a metal crystallite.

Consequently, structural, imaging, or other related experiments must be done to entangle the various combinations to chemisorption dimension D_C [48]. We must also know the site multiplicity of chemisorbed species, information readily obtained from infrared vibrational analysis. For instance, carbon monoxide chemisorbs on γ-Al_2O_3-dispersed Rh in two configurations, namely, as species $(CO)_2$ and as species CO [49]. (This phenomenon is not restricted to CO but has also been noticed for NO on alumina-dispersed molybdenum oxide [50].) Effectively, then, such observations imply a mixture of two chemisorption sites of different specific multiplicity; analysis of the chemisorption uptake by standard analytical techniques obviously would give a site average, weighted according to specific chemisorption site activity and to their relative occurrence on the support.

2.4.3   Metal Dispersion and Adsorbate Uptake: CO Chemisorption on Pt/γ-Al_2O_3

A need for minimizing noble metal loading such as Pt on high surface area γ-Al_2O_3 [44] under considerations of costs, and maximizing Pt metal dispersion under considerations of high catalytic activity, has engendered a large body of chemisorption studies of prototype system CO-Pt/γ-Al_2O_3. It is therefore challenging to apply the principles of chemisorption capacity m_C and its associated chemisorption dimension D_C to this situation. This pursuit is especially warranted as reports are extant that propose fundamental aspects of states of metal dispersion based, in part, on such CO chemisorption data.

Then, consider a representative study of a laboratory-made Pt/γ-Al_2O_3 catalyst [44] that chemisorbs CO in a single, "on-top" configuration [51]. The scaling plots of CO uptake versus Pt surface loading for data of a well-dispersed preparation are displayed in Fig. 2.16 by the open symbols, the catalyst having been hydrogen treated at 300 °C for 2 h prior to CO chemisorption (at room temperature). As both CO uptake and metal deposition data are not reduced to unit mass or volume, we reformulate Eq. (2.18) to

$$m_C \sim C^{D_c} \tag{2.19}$$

where C is the analytical metal loading. Least-squares slope s for the high-dispersed catalyst amounts to $s = D_C = 0.80 < 1$ (Fig. 2.16), which tell us that an increase in analytical Pt on the support does not lead to the same increase in CO chemisorption. Consequently, the scaling result does not support the conclusions (a) that the noble metal exists on the support as an entirely open, implicitly contiguous surface at atomic metal distribution—the δ phase [44]—(b) or that one Pt site chemisorbs just one CO molecule. For case (a), CO uptake ought to follow the dashed line of $D_C = 1$; we surmise that analytical preparation and catalyst treatment did not lead to perfectly open metal configurations. For situation (b), we estimate that it takes $1/0.80 = 1.25$ Pt sites per CO molecule, on average.

We now use Eq. (2.19) to describe CO chemisorption on sintered Pt states generated by subjecting the Pt/alumina catalysts to high temperatures under reducing and

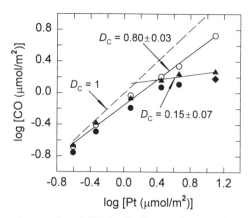

Fig. 2.16 Chemisorption uptake of CO by Pt dispersed on γ-Al$_2$O$_3$. The straight line of slope $D_C = 0.80$ [Eq. (2.19)] represents the least-squares linear regression for a highly Pt dispersed, nonsintered Pt/γ-Al$_2$O$_3$ catalyst ($\circ$). The linear regression line $D_C = 0.15$ represents the rate of CO uptake at high Pt loading for a catalysts sintered in O$_2$ at 500 °C for 5 h ($\blacktriangle$), in H$_2$ at 500 °C for 2 h ($\bullet$), and in H$_2$ at 750 °C for 2 h ($\blacklozenge$). The dashed line simulates $D_C = 1$ of perfect Pt dispersion and 1 : 1 CO site occupancy. [Reprinted with permission from H. C. Yao, M. Sieg, and H. K. Plummer, Jr., *J. Catal.* **59**, 365 (1979), Fig. 1B, Academic Press, Inc.]

oxidizing atmospheres prior to CO uptake [44] under the conditions enumerated in the legend of Fig. 2.16. Note that the effects of metal sintering (reduction of open Pt surface) by these harsh catalyst treatments are evident from lower CO uptakes than observed with the well-dispersed specimen. However, the increase of CO chemisorption uptake with analytical Pt concentration, as obtained from chemisorption dimension D_C (the slope of the log–log plots), remains essentially the same for high- and low-dispersed regimes—except that there is a cross-over point to near-zero slope, $s = D_C = 0.15$, for Pt loadings exceeding about 2 μmol/m^2 (BET). We shall now see that numerical value $D_C = 0.15$ does not agree with any reasonable scheme of *unrestricted* CO uptake by the metal but is an artifact caused by the preparatory method.

Consider that regardless of the shape of the metal crystallites of the sintered catalysts, increasing amounts of noble metal concentrations C on the alumina support must cause an increase in physical metal surface area S in one way or the other. Although we do not know the detailed Pt particle size distribution, we can nevertheless bracket the actual situation with the help of Eq. (2.19): We assume, on one hand, that the (average) *number density* of Pt crystallites (number of atoms/BET support area) is the determining factor for growth of total metal surface S with metal concentration C; hence $S \sim C^1$ or $D_C = 1$. Therefore, adding more metal increasingly fills empty support area. This outcome yields the largest feasible increase of open metal site generation under preparatively increasing metal loading (see the dashed line in Fig. 2.16).

On the other hand, we assume that the (average) *size* of the Pt crystallites is the factor for growth of total metal surface S under increasing metal loading; hence, $S \sim C^{2/3}$, $D_C = 0.67$. Therefore, we grow larger crystallites by stacking up on crystallites already at an alumina site, which yields a lower bound for the increase of CO chemisorption sites under preparatively increasing metal loading. Well, neither limit $D_C = 1$ nor $D_C = 0.67$ fits the experimental outcome of $D_C = 0.15$, even remotely.

The question then arises: Why does the increase in analytical Pt above 2 μmol/m^2 (BET) no longer lead to an increase of Pt sites able to chemisorb CO? Unless we involve the unlikely situation that only the corners of the large Pt crystals chemisorb CO, the simplest explanation for the observed $D_C = 0.15$ implies *blockage* of Pt chemisorption sites by metallic plugging of a pore entrance and/or metal filling of a pore [52], brought about by the reductive high-temperature sintering treatments of the lesser dispersed catalyst specimen at higher metal loading [44]. Indeed, a further prediction comes to mind: The high-temperature treated, high-metal-loading catalyst specimen become conducting by interior pore-occupancy percolation [53].

We therefore conclude that the apparent break at about 2 μmol/m^2 (BET) in the log–log plot of Fig. 2.16 does not support the proposed transition between a two-phase system of dispersed and crystalline Pt configurations [44]. This finding urges us to screen chemisorption data on the basis of chemisorption dimension D_C before what are actually preparatory-induced manifestations and effects are assigned to fundamental aspects of catalyst metal dispersion [54].

2.4.4   A Summary

Use of chemisorption scaling relations, such as chemisorption capacity m_C with particle radius (or diameter) R, offers useful information on the distribution of chemisorption sites under certain conditions and caveats. There is little extra work needed; chemisorption-catalyst loading data are simply graphed log–log [Eqs. (2.17)–(2.19); Figs 2.14 and 2.16]. If numerical information about the diameters of the active agent is not available, it is nevertheless possible to estimate limiting value of chemisorption dimension D_C from reported metal concentration or dispersion data.

As chemisorption usually entails fewer *active* than *total* sites, we ought to find $D_C < 2$ under Eq. (2.18) or (2.19). A range of $1 < D_C < 2$ may signify a fractal ridge structure of chemisorption sites (Fig. 2.15) or a distribution of Euclidian crystallite surface, corner, and edge sites. Once D_C drops below unity, active sites are perhaps fragmented into sets of disconnected chemisorption sites [37], conceivably on crystallite edges or on noncontiguous small metal clusters. We have seen that condition $D_C < 1$ can also be rationalized by multiple-site effects; a further example of this is presented in Chapter 4. Finally, scaling by Eq. (2.19) may afford an easy and quantitative means for ascertaining the reality of predictions concerning the structural nature of active chemisorption sites.

The concept of chemisorption capacity m_C and its associated chemisorption dimension D_C reaches further than heterogeneous catalysis. Recall that chemisorption signifies strong or specific adsorbent–adsorbate interactions (see Section 2.4.1). Hence, any irreversible tiling of an adsorbent with an adsorbate at ordinary temperatures may follow scaling relations Eq. (2.18) or (2.19). We expect that the obtained D_C describe scaling of *active* bonding or reaction sites with the analytical sites of the adsorbent, rather than scaling of its surface fractality [45, 48].

NOTES AND REFERENCES

[1] D. Avnir, D. Farin, and P. Pfeifer, Chemistry in noninteger dimensions between two and three. II. Fractal surfaces of adsorbents, *J. Chem. Phys.* **79,** 3566 (1983).

[2] W. Curtis Conner and C. O. Bennett, Are the pore and surface morphologies of real catalysts fractal?, *J. Chem. Soc. Faraday Trans.* **89,** 4109 (1993). An issue is made in distinguishing what the paper calls "ideal" fractals—the mathematical limit structures or attractors—from the "real" fractals found in catalysis. This is a priori a live concern that extends beyond fractality in catalysis (see section 8.3 of this book). Yet, it is frustrating to comment on this paper as the authors pour out the baby with the bath water; not many reported results of surface–pore analysis are considered valid. However, it is most unfortunate that the obvious usefulness and desirability of challenging fractal concepts is seriously undermined here by a misreading of basic aspects of fractal theory. For instance, Conner and Bennett steadfastly maintain that self-similarity makes fractals. This is wrong (the inverse relation *may* be true).

[3] In earlier versions of the manuscript, I called this "simulation", according to common usage. But this is wrong: A simulation means "to feign to have what one hasn't"; see *J. Baudrillard, Simulations* (Semiotext[e], 1983), p. 5. And as the laboratory data of Fig. 2.2 indicate, we have it!

[4] D. K. Ludlow and T. P. Moberg, Technique for determination of surface fractal dimension using a dynamic flow adsorption instrument, *Analyt. Instrum.* **19,** 113 (1990). As a reminder, NO_x stands for the most abundant offensive oxides of nitrogen, NO and NO_2. The reference gives no indication how the "fabric filter" was prepared (Manufacturer: Owens Corning Fiberglass)—its fractal surface dimension may arise from a highly crumpled interior material that practically fills a volume. (Conceivably, this causes non-Newtonian flow patterns, their tortuosity further increasing the efficiency of the physical trapping process.) On the other hand, an ordinary filter of $d = 2$ under straight flow-through pattern such as a honeycomb-shaped automotive converter, cannot offer these advantages. All told, Nature has since long developed fractal filtering, for instance, in plant configurations to efficiently sweep wind-blown pollen, in lung alvioli and in gills for optimum oxygen uptake.

[5] For a report addressing the extrapolation of the isotherm adsorption data to monolayer coverage (also including a crash course on the BET method), see J. J. Fripiat, L. Gatineau, and H. Van Damme, Multilayer physical adsorption on fractal surfaces, *Langmuir* **2,** 562 (1986).

[6] Strong objections were raised against the tiling method; see J. M. Drake, P. Levitz, and J. Klafter, Molecular adsorption on porous silica gels from binary solutions, *Isr. J. Chem.* **31,** 135 (1991). For opposing comments, see P. Pfeifer, D. Avnir, and D. Farin,

Complex surface geometry in nano-structure solids: Fractal versus Bernal-type models, in *Large-scale molecular systems,* W. Gans, A. Blumen, and A. Amann, Eds. (Plenum, New York, 1991), NATO Adv. Study Inst. **B258,** 215.

[7] Factor $\eta(\varepsilon)$ has been added to account for narrow Gaussian-distributed (or nearly so) surface irregularities. In other words, classical surface area S may depend *weakly* on linear size ε of the adsorbate.

[8] P. Pfeifer and D. Avnir, Chemistry in noninteger dimensions between two and three. I. Fractal theory of heterogeneous surfaces, *J. Chem. Phys.* **79,** 3558 (1983). The method appears somewhat cumbersome and lengthy, however.

[9] The profile of the Koch surface may be generated as follows: (a) Draw a straight, horizontal line segment of unit length (*initiator*). (b) Replace its middle third by an upside-down letter V, Λ, with each of its two branches having length $\frac{1}{3}$ and enfolding angle $60°$, all four pieces closing up precisely: This gives the *generator* (of length $1 - \frac{1}{3} + \frac{1}{3} + \frac{1}{3} = \frac{4}{3}$). (c) Iterate the preceding operation on *each* line segment. From Eq. (1.1): The number of line segments $\mathcal{N}(\varepsilon)$, each of length $\varepsilon = (\frac{1}{3})^n$, needed to cover stage n of the iteration, is $\mathcal{N}[(\frac{1}{3})^n] = 4^n = [(\frac{1}{3})^n]^{-d}$, $n = 1,2, \ldots$ It shows that $d = \log 4/\log 3 \approx 1.26$. This (in the *limit* $n \Rightarrow \infty$, $\varepsilon \Rightarrow 0$) is the Koch curve (see [2], Section 1.1).

[10] A look into Glossary 1 under *scaling* may be helpful in clarifying this very general procedure.

[11] See under *self-affine*, Glossary 1.

[12] D. Avnir, D. Farin, and P. Pfeifer, Surface geometric irregularities of particle materials: The fractal approach, *J. Coll. Interface Sci.* **103,** 112 (1985).

[13] Evidently, adsorbent radii R must be much larger than adsorbate size ε but lie substantially below the radius corresponding to the volume of a fractionated adsorbent sample. It appears that these two conditions can, generally, be met (see [8]).

[14] P. Pfeifer, Y. J. Wu, M. W. Cole, and J. Krim, Multilayer adsorption on a fractally rough surface, *Phys. Rev. Lett.* **62,** 1997 (1989). Related theoretical treatments, leading to the same theory, are given by Y. Yin, Adsorption isotherm on fractally porous materials, *Langmuir* **7,** 216 (1991); D. Avnir and M. Jaroniec, An isotherm equation for adsorption on fractal surfaces of heterogeneous porous materials, *Langmuir* **5,** 1431 (1991); F. Meng, J. R. Schlup, and L. T. Fan, Fractal analysis of polymeric and particulate titania aerogels by adsorption, *Chem. Mater.* **9,** 2459 (1997).

[15] K. Kaneko, M. Sato, T. Suzuki, Y. Fujiwara, K. Nishikawa, and M. Jaroniec, Surface fractal dimension of microporous carbon fibres by nitrogen adsorption, *J. Chem. Soc. Faraday Trans.* **87,** 179 (1991).

[16] J. M. Drake, L. N. Yacullo, P. Levitz, and J. Klafter, Nitrogen adsorption on porous silica: model-dependent analysis, *J. Phys. Chem.* **98,** 380 (1994).

[17] Consider a classical surface, where exponent $3 - d = 1$: Variations of ratio volume/ surface (or mass/surface) follow R^1, where R is a radius. For a surface fractal, variations of the ratio of volume/surface follow R^β, $\beta < 1$. Consequently, surface area of a surface fractal grows faster with volume or mass than for an object with a classical surface.

[18] A. V. Neimark, Thermodynamic method for calculating surface fractal dimension, *JETP Lett.* **51,** 607 (1990); A. V. Neimark, E. Robens, and K. K. Unger, Berechnung der Fraktaldimension einiger poröser Feststoffe aus der Stickstoff-Adsorptions-isotherme (Calculation of the fractal dimension of several porous solids from nitrogen adsorption isotherms), *Z. Phys. Chem.* **187,** 265 (1994).

[19] B. Sulikowsli, The fractal dimension in molecular sieves: Synthetic faujasite and related solids, *J. Phys. Chem.* **97,** 1420 (1993). Fractal analysis used Eq. (1.1) modified to $\mathcal{N} \sim V_m{}^{-d/3}$, where V_m is the molar adsorbate volume.

[20] Sulikowsli [19] correctly emphasizes that microporosity per se is not sufficient to engender fractal surface characteristics. (Remember: finer and finer detail.)

[21] E. Ignatzek, P. J. Plath, and U. Hünsdorf, The fractal character of zeolites. Part I: The fractal dimension of cobalt(ii)phthalocyanine loaded faujasite, *Z. phys. Chem.* (Leipzig) **268,** 859 (1987). The work involves inverse surface tiling of zeolite-incorporated phthalocyanines. However, the experimental scatter of the few data points is too large to make a credible statement of surface-size scaling behavior of the incorporated complexes; no error calculations (besides correlation coefficients over four data points) are given to better sustain the offered arguments and rationalizations.

[22] W. I. Friesen and R. J. Mikula, Fractal dimensions of coal particles, *J. Coll. Interface Sci.* **120,** 263 (1987).

[23] Cumulative pore volume V designates the momentary volume, from adding all differential pore volumes of pore radii ρ sampled successively from smaller and smaller pores at higher and higher intrusion pressures P. It scales with pore radius as $V(\rho) \sim \rho^{3-d}$. The pore surface distribution is obtained from $-dV/d\rho$, the minus sign rendering the differential quotient positive.

[24] Pressure P and pore radius ρ are related by the Washburn equation $P = (2\sigma/\rho) \cos \Theta$ in terms of interfacial tension σ and contact angle Θ. Inserting differentials $dP = 2\sigma \cos \Theta$ $[d(1/\rho)] \sim -(1/\rho^2)d\rho$ and $P \sim 1/\rho$ into Eq. (2.8) yields $dV/dP \sim -(dV/d\rho)\rho^2 \sim P^{d-4} \sim \rho^{4-d}$, or $-dV/d\rho \sim \rho^{2-d}$ [Eq. (2.7)].

[25] H. Spindler and M. Kraft, Fractal analysis of pores, *Catal. Today* **3,** 395 (1988).

[26] H. Spindler, W.-G. Ackermann, and M. Kraft, Die fraktale Dimension als Katalysatorkenngrösse. 2. Fraktale Beschreibung von Oberflächen und Porenvolumina, (The fractal dimension as characteristic quantity. 2. Fractal description of surfaces and pore volumes), *Z. phys. Chem.* (Leipzig) **269,** 1233 (1988).

[27] D. Avnir, D. Farin, and P. Pfeifer, A discussion of some aspects of surface fractality and of its determination, *New J. Chem.* **16,** 439 (1992).

[28] P. Pfeifer, G. P. Johnston, R. Deshpande, D. M. Smith, and A. J. Hurd, Structure of porous solids from preadsorbed films, *Langmuir* **7,** 2833 (1991).

[29] This ratio relates to the vapor pressures P (relative to saturation pressure P_0) of film preloading, not of the nitrogen.

[30] Preloaded film volume V is determined gravimetrically, whereas preloaded film surface area A is determined by subsequent standard BET analysis using N_2. The thickness of the preloaded film is estimated from Eqs. (2.9b) and (2.10): $A = dV/dz = (3 - d)z^{2-d}$, yielding $A = (3 - d)V/z$ or $z = (3 - d)V/A$.

[31] N. N. Jovanović, V. R. Nikolić, T. Novaković, and A. Terlecki-Baričević, Effect of rapid thermal decomposition of gibbsite on the fractal dimension of product surfaces, *J. Serb. Chem. Soc.* **58,** 229 (1993).

[32] A long, flexible linear polymer chain molecule embedded in Euclidian spaces of dimension $E \geq 2$ has mass fractal dimension (a) $d = \frac{5}{3}$ of self-avoiding random walk if swollen in a "good" solvent (prevalent repulsive interlink forces) and (b) $d = 2$ of ordinary random walk in a "theta" solvent (balanced repulsive-attractive interlink forces). See P.-G. de Gennes, Adsorption de polmères linéaires flexibles sur une surface

fractale, Adsorption of flexible linear polymers on a fractal surface, *C. R. Acad. Sci. Paris* **299**, Sér. II, 913 (1984).

[33] P. Pfeifer, *Interaction of fractals with fractals: Adsorption of polystyrene on porous Al_2O_3,* in *Fractals in Physics,* I. Pietronero and E. Tosatti, Eds. (North–Holland, Amsterdam, 1986), p. 47.

[34] Upon adsorption on a flat ($d = 2$) surface, the polymer chain is predicted to uncoil, forming a monolayer of monomers because of surface-monomer attraction and monomer-monomer repulsion; see [32].

[35] The absorbent fractal resembles portions of a diffusion-limited cluster aggregate ($d \approx 1.7$), a very real and important concept perhaps not for surface phenomena of our particular concern here but certainly in other applications of fractals in chemistry. This subject will be taken up in Chapter 3.

[36] A highly porous, $d \approx 3$ adsorbent keeps adsorbed polymer chains in their solution configuration (see [32, 33]). Incidentally, this justifies the covering method of high-surface γ-Al_2O_3 by randomly coiled styrene–methyl methacrylate copolymers displayed in Fig. 2.2.

[37] D. Avnir and D. Farin, Fractal scaling laws in heterogeneous chemistry: Part I: Adsorptions, chemisorptions, and interactions between adsorbates, *New J. Chem.* **14,** 197 (1990).

[38] A. Venkatraman, L. T. Fan, and W. P. Walawender, The influence of the temperature of calcination on the surface fractal dimensions of $Ca(OH)_2$-derived sorbents, *J. Coll. Interface Sci.* **182,** 578 (1996).

[39] K. Oshida, K. Kogiso, K. Matsubayashi, K. Takeuchi, S. Kobayashi, M. Endo, M. S. Dresselhaus, and G. Dresselhaus, Analysis of pore structure of activated carbon fibers using high resolution transmission electron microscopy and image processing, *J. Mater. Res.* **10,** 2507 (1995).

[40] H. S. Gandhi, H. C. Yao, H. K. Stepien, and M. Shelef, Evaluation of three-way catalysts. Part III. Formation of ammonia, its suppression by sulfur dioxide and reoxidation, *SAE Spec. Publ. SP-431* (1978), Inter-Ind. Emiss. Control Program 2 (IIEC-2) Prog. Rep. No. 4, 107.

[41] E. C. Su, W. G. Rothschild, and H. C. Yao, Carbon monoxide oxidation over platinum/ γ-alumina under high pressure, *J. Catal.* **118,** 111 (1989).

[42] R. M. Ziff and K. Fichthorn, Fractal clustering of reactants on a catalyst surface, *Phys. Rev. B* **34,** 2038 (1986). The model discounts surface diffusion.

[43] B. J. Kip, F. B. M. Duivenvoorden, D. C. Koningsberger, and R. Prins, Determination of metal particle size of highly dispersed Rh, Ir, and Pt catalysts by hydrogen chemisorption and EXAFS, *J. Am. Chem. Soc.* **108,** 5633 (1986).

[44] H. C. Yao, M. Sieg, and H. K. Plummer, Jr., Surface interactions in the Pt/γ-Al_2O_3 system, *J. Catal.* **59,** 365 (1979).

[45] D. Farin and D. Avnir, *The fractal nature of molecule–surface interactions and reactions,* in *The Fractal Approach to Heterogeneous Chemistry,* D. Avnir, Ed. (Wiley, Chichester, UK, 1989), p. 284.

[46] See p. 124 of [2], Section 1.1.

[47] P. Pfeifer, Katalysatoroberflächen, Makromoleküle und Kolloidaggregate: Fraktale Dimension als versteckte Symmetrie unregelmässiger Strukturen (Catalyst surfaces, macromolecules, and colloidal aggregates: Fractal dimension as hidden symmetry of irregular structures), *Chimia* **39,** 120 (1985).

[48] D. Avnir, Fractal geometry—A new approach to heterogeneous catalysis, *Chem. Ind.,* 16 Dec. 1991, 912.

[49] H. C. Yao and W. G. Rothschild, Infrared spectra of chemisorbed CO on $Rh/\gamma\text{-}Al_2O_3$: Site distributions and molecular mobility, *J. Chem. Phys.* **68,** 4774 (1978). The stretching fundamentals of two adjacent CO molecules [immobilized at room temperature against fast surface diffusion by their chemisorption site(s)] enter into vibrational resonance, splitting the CO stretch into a symmetric and antisymmetric stretching mode. In addition, two bending fundamentals of the four-atom configuration are generated [considering the fifth member, the adsorbent site(s), of infinite mass]. For adsorption on a *metallic* site (large crystallite), the antisymmetric mode is forbidden as its displacement vector lies parallel to the metal surface. Hence, the appearance of the doublet is a valid indication of *non*metallic character of the noble metal at the site (small particles). Interestingly, the dihedral angle between the two CO moieties varies with the oxidation state of the Rh. This may have implications for reaction selectivity and steering of this activated complex.

[50] H. C. Yao, W. G. Rothschild, and H. S. Gandhi, *Infrared spectra of NO chemisorbed on a* $Pt/MoO_3/Al_2O_3$ *catalyst,* in *Catalysis on the Energy Scene,* S. Kaliaguine and A. Mahay, Eds. (Elsevier, Amsterdam, 1984), p. 71; H. C. Yao and Walter G. Rothschild, Surface interaction in the $MoO_3/\gamma\text{-}Al_2O_3$ system. II. Effect of surface structure on NO chemisorption, in *Proceedings of the Fourth International Conference, The Chemistry and Uses of Molybdenum* (Climax Molybdenum Co., Ann Arbor, MI, 1982), p. 31.

[51] W. G. Rothschild and H. C. Yao, Surface interaction in the $Pt/\gamma\text{-}Al_2O_3$ system. II. Dynamics of CO chemisorption by infrared fluctuation spectroscopy, *J. Chem. Phys.* **74,** 4186 (1981).

[52] W. G. Rothschild, Fractals in heterogeneous catalysis, *Catal. Rev. Sci. Eng.* **33,** 71 (1991).

[53] I first imagined that it was me who had found something here, but later learned that the originality of the concepts of pore blocking and percolative conductivity had been predicted a long time ago: See J. M. Hammersley, *Monte Carlo Methods* (Wiley, New York, 1964), pp. 134–135. For another paper on catalyst pore blocking, refer to G. A. Simons and A. R. Garman, Small pore closure and deactivation of the limestone sulfation reaction, *AIChE J.* **32,** 1491 (1986).

[54] For two more examples, see Figs. 7 and 8 in [52].

3

FRACTALITY IN AGGREGATION AND GROWTH

3.1 FRACTAL ASPECTS OF AGGLOMERATE STRUCTURES

So far, the text has presented effects of surface fractality on adsorptivity and permeability without emphasis on modes of materials preparation. Now we ought to consider the various processes and mechanisms that actually generate fractal adsorbents. Such an undertaking is not only relevant to the praxis, for instance in preparing adsorbents with desired fractal characteristics, but also highly instructive from a scientific point of view.

We had discussed mechanisms of pore generation in initially nonporous materials by forced ejection of gaseous components, such as activated alumina from loss of crystal water of gibbsite (Section 2.3.4) and activated carbons from loss of hydrocarbons of tars and coals (Section 2.3.1). Under milder, persistent erosion regimes these processes fall under the headings of leaching, drying, desiccation, plasticizer loss, and drainage. (Next time you walk over a parking lot, notice the shape of the cracks in the asphalt.) All these material deterioration mechanisms are of prime interest to material scientists, geologists, and engineers, but some of them pertain directly to chemistry and will occupy us later.

Another process of forming fractally distributed particles is by fragmentation of larger, originally Euclidian-dimensioned bodies. While again of obvious interest to geologists and engineers, fragmentation phenomena are also of significant relevance in the food processing and packaging industry (Section 6.1.1).

Now, however, we will discuss another approach of generating fractal objects that is of special concern to chemists, namely formation of fractal objects by aggregation of smaller, nearly identically sized particles. Because the mechanisms contain random elements, we can expect three main structural phenomena arising in the aggregated objects: (a) Mass fractality, because the process often leads to more or less diaphanous aggregates, (b) fractal pore domains consisting of interstitial void elements of wider range of linear size, and (c) surface fractality from the interfaces between pores and mass regions. Although it is commonly difficult to differentiate between the resulting pore and surface fractal [1], we need to explore the fractal distributions [Eq. (1.1)] of mass, surface, and pores and their effects on selected chemical processes.

In general, what type of particular mechanisms leading to fractal aggregates would be of interest to chemists and of applicability in their work? Two widely different processes of considerable scope and utility come to mind, namely, gel formation by polymerization reactions of polyvalent monomers, and diffusion-limited aggregation (DLA). Mechanistically, the first process generates large molecular networks by definite and well-understood chemical reaction steps carried out in the laboratory. The second process grows large clusters by purely random aggregation of smaller entities and is, a priori, a computer model but its outcome mirrors a large variety of physical objects and chemical dynamics.

3.1.1 Application of Gel Formation: Size-Exclusion Chromatography

Size exclusion chromatography—a special technique in the set of high performance liquid chromatography (HPLC)—is perhaps not of great interest to most chemists. It came to my attention recently when glancing over the polymer literature, motivated by a standing attachment to this field of chemistry [2]. Indeed, chain growth of polyfunctional monomers to yield gelatinous agglomerates underpins fabrication of versatile size-exclusion column materials.

But why are gels relevant? A gel represents a peculiar state of matter of polydispersed, branched (cross-linked) polymer chains, a "two-component system of semisolid nature, rich in liquid" [3]. These liquid regions, called sol, furnish the volume elements effective for separation into categories by *physical size* of the mixture of molecular solutes that are put through the apparatus. To see this in greater detail, we now briefly discuss appropriate aspects of gel formation and percolation.

Consider a progressing polymerization reaction based on a *poly*functional monomer, via polymer–monomer and polymer–polymer growth and chain-termination steps. The reaction yields increasingly larger size, cross-linked clusters within a wide size distribution of smaller clusters and nonreacted monomer. Because of the polyfunctional monomer, the polymeric chains contain intrachain loops (a few) and interchain branches (very many), with some proportion of dangling bonds. Significantly, at some stage of the reaction, a *critical* system state of chain length and chain branching is reached such that one large polymer cluster—that is, a molecule of *connected* monomer units—is able to reach through the entire reaction vessel. This state is aptly termed (bond) "percolation" and the (in principle, infinitely large) percolation cluster is called "incipient gel."

Now, the incipient gel is of interest to us because it is a statistically self-similar fractal of dimension $d \approx 2.5$ if reactions take place, as usual, within volumetric containers or vats. Note that the value of this dimension is inferior to that of a volume ($d = 3$) but exceeds those of a regular surface ($d = 2$) and of polymeric chain segments ($d = 1$).

Pushing the extent of the polymerization reactions further and further, remaining individual reactive finite clusters (lattice animals) as well as nonreacted monomer increasingly combine with the percolation cluster. Thereby, the reactions form larger and larger regions of a *cross-linked polymer network* that are no longer statistically self-similar under one fractal dimension. (The limit structure would be a three-dimensional polymer network.) This statistically self-similar fractal set of finite clusters and nonreacted monomer (the sol), scaling from the size of a monomer unit up to its longest chains, does the size separation of the chromatographic apparatus.

The fractal medium is characterized conveniently by a problem-related mass (M)–distance (R) scaling relation with associated fractal dimension d [see Eq. (2.2), and *scaling* in Glossary 1],

$$M \sim R^d \tag{3.1}$$

For our present purpose, namely, reaction of multifunctional monomer molecules to yield polydispersed, branched polymers, mass M refers to number-average mass $M_n \sim \int Mf(M)dM$, to weight-average mass $M_w \sim \int M^2 f(M)dM$, or to z-average mass $M_z \sim \int M^3 f(M)dM$, all under normalized weighting distribution $f(M)$. The concomitant (average) radii, R_n, R_w, R_z reflect the effects of the corresponding mass weighting; clearly, radius R_z characterizes the largest polymer clusters of the sol—indeed those of greatest interest here. Designating from now on R_z by L, we realize that L serves as the *upper* molecular separation limit: Distributions of solute molecules of radii R *exceeding* L are excluded from the sol, forming therefore the first-appearing elution peak upon rapid passage through the void spaces of the column.

The experimental elution data are generally evaluated in terms of the so-called partition (or distribution) coefficient K_d. It is defined by [4]

$$K_d = (v_e - v_o)/(v_t - v_o) = (v_e - v_o)/v_p \tag{3.2}$$

where the subscripts identify the various volume elements (v), namely, elution v_e, interstitial (or void) v_o, total v_t, and internal pore v_p. Coefficient K_d then designates the fraction of the internal pore volume accessible to the solute molecules. Its limiting values are 1 and 0, derived as follows: (a) For solute diameters $R \ll L$, recalling that $R_z \equiv L$ is the size of the largest pore in the sol, elution distributes solute molecules randomly between sol and interstitial (void) volumes: $v_e = v_o + v_p$, hence $K_d = 1$. (b) For solute diameters $R \geq L$, size exclusion elutes solute molecules through the void spaces. Thus $v_e = v_o$ and $K_d = 0$.

To write an appropriate and useful expression for K_d comprising fractal aspects of the sole pore distributions [5], we fall back on previously discussed fractal theory of porosity, setting

$$K_d = 1 - (R/L)^{3-d} \tag{3.3}$$

Indeed, Eq. (3.3) contains the following pertinent fractal aspects:

1. The relation includes a power law distribution $\mathcal{N}(R) \sim R^{-d}$ of pore radii (or diameters) under fractal dimension d [see Eq. (1.1)].
2. Term $(R/L)^{3-d}$ represents a fractal pore volume–radius dependence, $V(R/L) \sim [\mathcal{N}(R/L)](R/L)^3$, relative to largest pore size L [6] [see also Eq. (2.10)].
3. The complement of $(R/L)^{3-d}$ to unity, that is, Eq. (3.3), indeed covers the entire range of size-exclusion effects from $K_d \Rightarrow 0$ for $R \Rightarrow L$ to $K_d \Rightarrow 1$ for $R \ll L$.

Scaling relation Eq. (3.3) is evaluated by plotting $\log(R)$ versus $\log(1 - K_d)$,

$$\log(R) = [1/(3-d)]\log(1 - K_d) + \log(L) \qquad (3.4)$$

Hence, the intercept at $K_d \Rightarrow 0$ yields the largest pore radius L of the packing, whereas the slope gives the inverse excess dimension $3 - d$ (see Section 2.3.1). Conveniently, polymer radius R is expressed in terms of molecular weight, taking proper account of solvent properties, by using Eq. (2.13) or inverting Eq. (3.1): Here $R \sim M^{1/d} \equiv M^{\upsilon}$ [7]. We then substitute in Eq. (3.4) $\log(R)$ by $\upsilon \log(M)$, which modifies factor $1/(3-d)$ to $1/[(3-d)\upsilon]$, term $\log(R)$ to $\log(M)$, and $\log(L)$ to $\log(M_L)$, with $M_L = L^{1/\upsilon}$ or $L = M_L^{\upsilon}$. The quantity M_L is the molecular weight of the *smallest* solute molecule *excluded* from entering the sol pore structure [8].

All parameters of Eq. (3.4) are accessible by experiment: Distribution coefficient K_d from elution data on the (calibrated) column, molecular weight M by the ultracentrifuge sedimentation techniques, and critical exponent υ from long-established theory or computer calculations.

But does this all work? Well, yes and no. For example: (a) Aqueous dextran on μ-Bondagel-E500, a silica-based pore structure packing (with minimized adsorptive properties to avoid specific adsorbent–adsorbate interaction), yielded dimension $d = 2.67$ ($\upsilon = \frac{3}{8}$) over range $200 < M < 3.16 \times 10^5$ with extrapolation to $M_L = 6.8 \times 10^5$. (b) Water-soluble globular proteins on HPLC gels TSK4000PW and 5000PW gave $d = 2.45$, $L = 14.3$ Å and $d = 2.36$, $L = 209$ Å, respectively. Evidently, the pore distribution in these systems is fractal—pore surface dimensions greatly exceeding classical value $d = 2$. (c) Standard water-soluble proteins on column gels Sepharose and Sephacryl turned out to yield sigmoid-shaped log–log plots, perhaps reflecting the influence of more than one fractal dimension, each governing a certain partial range of molecular weight. This explanation agrees with the well-known heterogeneity of the pore distributions of these gels. At present, however, such assumption of multifractality remains speculative (see also Section 2.3.2).

Part of the difficulties of obtaining a generally good fit of theory to experiment rests, in all likelihood, with a neglect of interaction forces between adsorbent and adsorbate (see also Section 2.2). A neglect of adsorbate–adsorbent forces seems particularly relevant to gels based on silica, which carries active oxygen and hydroxyl groups. Attempts have been made to quantitatively ascertain and separate out such secondary effects, using solvents of different column reactivity (polar, nonpolar) [9]. However, a discussion of such a specialized topic has no place in this book, except to point out that fractal aspects in size-exclusion chromatography are more complex than would appear from simple theory.

3.1.2 Application of Diffusion-Limited Aggregation

We turn now to a mechanism of cluster growth by diffusion-limited aggregation (DLA). In the simplest case, this involves randomly drifting particles within a system containing a preexistent nucleus and their instantaneous, irreversible attachment upon a particle–nucleus encounter. Ideas of what to expect from the process are best obtained from computer-generated random walk schemes: It turns out that they are pretty close to many experimental outcomes.

In brief, particles placed at the rim of a coordinate system embedded in some space and containing a nucleus, are set off to diffuse independently (one particle after the other) by random walk over the lattice points. Once a particle has reached the nucleus at nearest-neighbor distance, it irreversibly and instantaneously adheres to it [10]. (Particle diffusion is therefore the slowest reaction step; hence the descriptor "diffusion limited.") Eventually, a branching, wispy-looking random structure of connected nearest-neighbor sites emerges within the lattice, displaying many "hairy arms" radiating out in an irregular fashion from the central region of the initial nucleus. The shape of the structure of the central region remains relatively constant because the reach of the outer branches of the growing aggregate efficiently nets approaching random walkers, preventing most of them from attaining the interior. As typical of a fractal, the aggregate has no natural length scale within the resolution limits ranging from the size of the individual particle to distances well below its radius of gyration.

Regarding shape and dimension of the underlying lattice, a square grid is conveniently used but triangular and hexagonal coordinates are also appropriate; the particular form is generally unimportant. On the other hand, the dimension of the grid space matters greatly: This effect should come as no surprise as higher dimensioned lattices force the random walk to explore more coordinate directions.

Fractal properties are conveniently expressed by a now familiar mass–distance scaling relation

$$N \sim R^d \tag{3.5}$$

where N signifies the number of particles within radius of gyration $R(N)$ and d is the fractal dimension of the aggregate. Equation (3.5) is just an equivalent way of formulating the basic scaling relation Eq. (3.1). From computer models, the fractal dimension d of DLA is found to be close to $\frac{5}{3} = 1.7$ for the two-dimensional square lattice and approaching $\frac{5}{2} = 2.5$ for a three dimensional lattice; d is given by general scaling relation $d = (\frac{5}{6})E$, where $E = 2$ (square grid) or $E = 3$ (cubic lattice) is the Euclidian space dimension [11–13]. Bear in mind that the walk takes place on a lattice of a perfectly normal Euclidian space but that the final, global aggregate object of the DLA process is a fractal of noninteger dimension $d < E$: Not all available sites of E are reached by the walkers (see above).

Although the computed fractal dimension $d \approx 2.5$ of the DLA cluster embedded in $E = 3$ space is seen to be rather close to representative values of dimensions of actual pore and surface fractals discussed in Chapter 2, DLA must be related to real-world fractal agglomeration structures with great care. First, if simplifying assumptions underlying the model, in particular a perfect "sticking efficiency," are lifted by allowing several "collision" attempts before a diffusing particle adheres irreversibly to the cluster, the form of the global cluster is much less irregularly branched: In fact, it approaches the underlying grid symmetry (square, triangular, . . .) with increasing sticking attempt frequency. Second, it turned out that sufficiently large lattices generate the same effect as multiple sticking attempts in much smaller lattices. This unexpected observation showed up under large-scale computer modeling.

A more complicated DLA mechanism involves diffusion-limited cluster–cluster aggregation [14]. Instead of individual particles diffusing randomly toward a growing nucleus, individual clusters on a lattice are modeled to undergo translational steps by one lattice unit in a random direction. Following each step, nearest-neighbor contact with any other cluster is probed. If affirmative, the two clusters are assumed to stick together instantaneously and irreversibly to a larger cluster. The process iterates on all clusters so that, in the end, a single large cluster remains [15]. Cluster–cluster DLA on three-dimensional lattices—those of principal interest to us—yields fractal dimension $d = 1.7$–1.8. Note that this value is not only significantly smaller than the corresponding dimension for three-dimensional particle–cluster DLA, namely, $d \approx 2.5$, but does not even attain the space-filling capacity of a classical surface. Why? Simply because a diffusing cluster aggregate is even less likely to penetrate into the central regions of another cluster than a single diffusing particle. Consequently, cluster–cluster DLA leads to rather tenuous shapes.

Although newer results of computer-generated DLA processes have turned out to be more complicated than previously assumed [14, 16], the simple DLA models serve us well as long as we consider the *interior* regions of an agglomerate. Why? Because their growth has ceased (no further walkers are able to reach them). Hence, the more central portions of DLA clusters do not change materially under continuing agglomeration at their rim sections, and therefore stay in accordance with simple DLA theory (independence of lattice shape). Indeed, this seems to me to be an instance that actually *exploits* the finite range of linear scales of physical fractals. We shall see later that DLA is not the only example.

Fractal dimension d of a DLA computer-generated process is indeed a useful indicator for predicting and evaluating aggregation clusters encountered in the laboratory. Furthermore, DLA processes are present in the manufacture of aerogels, colloids, paper, coatings, magnetic tapes, and refractory catalyst supports. The DLA clusters are present in filler and pigmentation materials, used in flotation of minerals, and act as important participants in atmospheric pollution in the form of aerosols and particulates. Two examples under representative analytical methods were selected: silica aggregation studied by image analysis, and soot agglomeration characterized by light scattering.

3.1.2.1 Image Analysis of Silica Aggregation

Fractal silica aggregates, precisely because of their tenuous nonclassical geometry [17], have been used to increase the tensil strength of rubber—perhaps that fractally distributed interaction sites on polymer molecules are preferably cross-linked by a fractal filler as it simply affords improved matching within the polymeric network (see, e.g., Sections 2.3.5 and 3.1.1). Pertinent results [18] are displayed in Fig. 3.1. Figure 3.1(a) shows the structure of an agglomerate, from computer-generated diffusion-limited cluster–cluster aggregation with $d = 1.8$, in terms of planar projections (XY, YZ, ZX) for better viewing. Each disk represents an individual constituent cluster. Figure 3.1(b) displays two microscopic images of cluster–cluster aggregates of (acid-precipitated) silica, each of its spheroidal constituent clusters of diameter approximately equal to 250 Å. Figure 3.2 presents the corresponding log–log plot of aggregate mass versus its size [Eq. (3.5)], with size here signifying the average of the maximal width and height of a cluster projection (Feret's diameters [19]). A mass fractal dimension of $d = 1.51 \pm 0.09$ was calculated (unrestricted least-squares) from the presented data points [18]. It lies 17% below computer-predicted dimension $d = 1.8$ of cluster–cluster DLA.

Additional information on the fractal properties of the aggregates is obtainable from a scaling operation of the perimeter of their images [19, 20]; we now consider the aggregated object as a so-called boundary fractal—which emphasizes that different experimental procedures may probe different fractal characteristics of one and the same macroscopic object. The perimeter is measured by stepping around under step size ε, with ε varying between the experimental cutoffs—the diameter of a constituent cluster (lower limit) and Feret's diameter (upper limit). The slope of

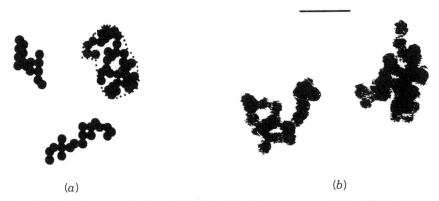

(a)   (b)

Fig. 3.1 (a) Planar projections XY, YZ, ZX of computer-modeled diffusion-limited cluster–cluster aggregation. The dotted segments demonstrate a boundary measurement of an aggregate by the step divider procedure, shown here for step size of about 400 Å. (b) Microscopic images of acid-precipitated silica aggregates composed of near-identical, spheroidal SiO$_2$ clusters of about 250-Å each. The length of the horizontal bar above the images corresponds to 1000 Å. [Reprinted with permission from A. Tuel, P. Dautry, H. Hommel, A. P. Legrand, and J. C. Morawski, *Prog. Coll. Polym. Sci.* **76**, 32 (1988), Fig. 5, Copyright Dr. Dietrich Steinkopff Verlag.]

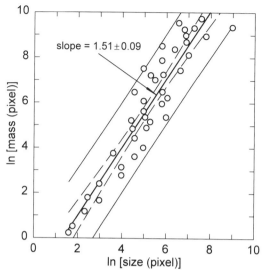

Fig. 3.2 Log–log (natural) mass–size scaling plot of an image analysis of diffusion-limited cluster–cluster aggregates of primary silica clusters of 250-Å diameter each [Fig. 3.1(*b*)]. The linear least-squares slope, its standard deviation, the 95% confidence limit (dashed curve), and prediction intervals correspond to the mass fractal dimension of all measured aggregates. [Reprinted with permission from A. Tuel, P. Dautry, H. Hommel, A. P. Legrand, and J. C. Morawski, *Prog. Coll. Polym. Sci.* **76**, 32 (1988), Fig. 4, Copyright Dr. Dietrich Steinkopff Verlag.]

$\log \mathcal{N}(\varepsilon)$—the number of segments of size ε to cover the perimeter—versus $\log(1/\varepsilon)$ then gives the fractal dimension of the boundary of the aggregate [see Eq. (1.1)]: The result is $d \approx 1.3$ [18]. We know well what this means: The perimeter is contorted (at finer and finer detail) relative to an ordinary boundary curve ($d = 1$).

3.1.2.2 Light Scattering Analysis of Soot Agglomerates and Processed Brown Coal

While it is not necessary to emphasize the importance and relevance of soot in industrial applications and environmental concerns to a readership of chemists, it may be of interest and relevance to search for fractality in soot, a tenuous material likely arising via DLA aggregation.

A versatile as well as comprehensive investigative method for evaluating fractality of soot is laser light scattering. It has two advantages: First, it affords in situ observation, which not only permits recording of light scattered from distant sources but also guarantees that shape and density of the soot are not altered by sample manipulation. Second, light scattering [21] effectively scales distances on the fractal object by recording the scattered intensity $I(q)$ as a function of a quantity called wave vector and frequently, but by no means consistently, designated by symbol q; quantity q is conveniently given in units of reciprocal centimeter (but this also is not universal practice).

The significance of q is best appreciated by its reciprocal quantity q^{-1}, which represents a resolution length of the scattering experiment. Then, by picking a series of values of wave vector q—readily done by viewing the scattered light under varying scattering angles θ—different length scales on the fractal object can be probed. Consequently, a specific value of quantity q^{-1} (or angle θ) corresponds to a specific length pertaining to the fractal object. Again, it is the same principle as discussed repeatedly in this text: A scaling operation of counting the number of (self-similar) portions that make up the object, as function of their linear extent.

The well-known relation between wave vector q, scattering angle θ, and radiation wavelength λ is

$$q = 4\pi\lambda^{-1} \sin(\theta/2) \tag{3.6}$$

neglecting (small) changes of the index of refraction. Hence, a rather large range of q values, and accordingly a wide scope of distances on the fractal, can thereby be explored by varying θ and λ. Let us then ascertain the type of fractal characteristics accessible through this technique by looking at the ranges of q^{-1}. If resolution length q^{-1} is larger than the size of the fractal—characterized by its radius of gyration r_G (see Section 2.3.5)—any fractal irregularities are washed out in the scattering experiment: A measurement of scattered intensity $I(q)$ effectively only gives a measure of radius of gyration r_G [22]:

$$I(q) \sim 1 - q^2\, r_G^2/3 \qquad q^{-1} > r_G \tag{3.7}$$

On the other hand, if resolution length obeys $r_{PRIM} < q^{-1} < r_G$, where r_{PRIM} is the radius of the smallest constituent (primary) cluster of the aggregate, the observations sample the mass–radius distribution [Eq. (3.1)] of the entire aggregate. The resulting dimension, $d = d_M$, obtains from relation

$$I(q) \sim q^{-d} \quad d = d_M \quad r_{PRIM} < q^{-1} < r_G \tag{3.8}$$

[Note that Eq. (3.8) may be derived from Eq. (2.17) by assigning $I(q)$ to *property* and q^{-1} to *scale*.] Finally, condition $q^{-1} < r_{PRIM}$ evidently probes irregularities inherent in the primary or constituent clusters of the total aggregate, leading to surface fractal dimension $d = d_S$ [23]

$$I(q) \sim q^{d-6} \qquad d = d_S, \quad q^{-1} < r_{PRIM} \tag{3.9}$$

(written here for three-dimensional embedding space). Hence, the following ranges of scattering angle θ are appropriate:

1. Determination of radius of gyration r_G via Eq. (3.7). Taking, say $r_G = 2000$ Å, would require sampling of the scattered radiation intensity $I(q)$ at $q^{-1} \geq 2000$ Å or $q \leq 0.0005$ Å^{-1}. With Ar laser line 5415 Å, we ought to select $\sin(\theta/2)$ inferior to $(1/4\pi) \times 0.0005 \times 5415 = 0.215$ or $\theta \leq 25°$ [see Eq. (3.6)].

2. Determination of a fractal dimension d from the slope of plots $\log[I(q)]$ versus $\log(q)$ (slope is negative) or versus $\log(1/q)$ (positive slope) over the maximal experimentally available range of q. Scattering angles θ must now exceed $25°$ in order to probe length scales $q^{-1} < r_G$. But which fractal dimension, $d = d_M$ for mass or $d = d_S$ for surface, do we measure? With natural limits d_S, $d_M \leq 3$, Eqs. (3.8) and (3.9) predict that the numerical value of the slope of a $\log[I(q)] - \log(1/q)$ plot exceeding 3 (falling below 3) would pertain to a surface (mass) fractal.

Experimental exploration of these principles has been reported for laser light scattered off soot particles in the combustion zone of an ethylene/air flame [22]. Typical values of $\log[I(q)]/\log(q) = -1.5$ infer that the soot aggregates constitute mass fractals [Eq (3.8)]. Notice that dimension $d_M = 1.5$ is below $d_M = 1.8$ predicted for diffusion-limited cluster–cluster aggregation in three-dimensional embedding space (Section 3.1.2) but equals the experimental outcome from precipitated silica aggregates (Section 3.1.2.1).

Soot is formed by aggregation of carbon particles in the combustion zone of a hydrocarbon/air flame, silica aggregates are formed by precipitation in a liquid medium (Section 3.1.2.1). Yet, both agglomeration processes are characterized by about the same fractal dimension. Hence, the assumption that they are generated by the same simple code—diffusion-limited cluster–cluster aggregation—is more than idle speculation (but proof it is not).

We conclude with a brief discussion of X-ray scattering of treated brown coal. Brown coal, the dull-looking, moisture-containing fossil that "did not quite make it"—in comparison to the beautifully glossy stuff dug up in Newcastle, the Ruhr, Kentucky, and other blessed places. However, this disadvantage has not discouraged chemists and chemical engineers from finding value-added industrial uses for brown coal. Such syntheses require good quality control and a precise characterization of the coal surface parameters as, upon processing, noncarbonized volatile fluids pass through its structure. Hence, questions on the nature and degree of porosity and surface irregularities need to be addressed.

Figure 3.3 shows a plot of $\log[I(q)]$ versus $\log(q)$ of an alkali-digested, solar-dried brown coal slurry [24]. Within range 0.35–10 mrad (0.02°–0.57°) of scattering angle θ under incident radiation of a Cu anode X-ray tube, the plot is linear. It yields a slope of $s = -3.4$ over a range of scattering angle of a factor of about 25. Relating $s = -3.4$ to the exponent of Eq. (3.8) yields mass fractal dimension $d_M = 3.4 > 3$, which does not make sense. Relating slope $s = -3.4$ to exponent $d_S - 6$ of Eq. (3.9) yields surface dimension $d_S = 2.6$, an acceptable value: The treated brown coal is probably a surface fractal.

3.1.3 Mass, Surface, Pore Fractal?

At the beginning of this chapter, the question of the difficulty of distinguishing between surface and pore fractals was raised. Perhaps this may not be an urgent point; knowing the fractal dimension of the object of interest is f ᵔ᷉uently all that is desirable

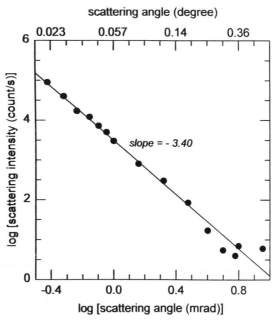

Fig. 3.3 Log–log (decimal) scaling plot of X-ray scattering intensity data for an alkali-digested brown coal slurry, probing its surface fractal dimension $d = 6 + (\text{slope} = -3.4) = 2.6$ [see Eq. (3.9)]. [Reproduced with permission from I. Snook and P. McMahon, *Langmuir* **9**, 2726 (1993), Fig. 1, Copyright 1993 American Chemical Society.]

and useful. On the other hand, we have seen above that light scattering techniques offer a possibility to sharpen the character of a fractal by isolating its different fractal properties [see Eqs. (3.8) and (3.9)].

But what does one do if the fractal is truly mixed, a composite mass, surface, pore fractal under different dimensions for each property? It would not only be instructive but also useful to establish a model predicting how the registered scattered intensity should scale with wave vector q under some combined exponent of, perhaps as much as three, different dimensions. The underlying formalism has been established [25]; undoubtedly, application ought to be seriously considered. Further, its theoretical-heuristic development is quite interesting and instructive.

As first step in the theory, the law between scattering intensity and wave vector is generalized to

$$I(q) \sim q^{f(d_M,\, d_S,\, d_P)} \tag{3.10}$$

where exponent $f(d_M, d_S, d_P)$ is assumed to be a linear function of mass, surface, and pore dimensions. In general notation ($d_M = \alpha_1$, and so on) we then write

$$f(\alpha_1, \alpha_2, \alpha_3) = a_0 + a_1\alpha_1 + a_2\alpha_2 + a_3\alpha_3 = a_0 + a_1 d_M + a_2 d_S + a_3 d_P \tag{3.11}$$

having to solve for the four coefficients a_i. To continue, we must first know (a) how a purely mass fractal scales with surface and pore area, (b) how a purely surface fractal scales with mass and pore area, and (c) how a purely pore fractal scales with mass and surface. The appropriate scaling relations of

$$\text{property} \sim \text{distance}^d \tag{3.12}$$

under property = mass (volume), or surface, or pore area, respectively, are collected in Table 3.1. This table also gives the range of validity of fractal dimension d under its property. Fragmented fractals such as the Cantor set (see *attractor,* Glossary 1) have been omitted.

To give some visual impressions, Fig. 3.4(*a*) shows a photograph I took of a piece of scorched earth; it demonstrates well a statistically self-similar purely *pore* fractal of some pore dimension d, with mass (the big clumps of ground around each fissure) scaling as R^3 and the surface—the interface between pore and mass [26]— scaling with the same dimension d as the pores (see also Table 3.1). Now, imagine a downpour, water collecting quickly in the pores, and you have a purely *mass* fractal of dimension d, pores scaling as R^3, and the interface again scaling as R^d. Figure 3.4(*b*) represents a microphotograph of a leaf section of 2.3×1.8-mm^2 size, its pattern forming another (purely) pore or mass fractal depending on viewing it as the negative (pore) or the positive (mass) image. Finally, Fig. 3.5 shows scanning electron microscopic images of a catalyst particle, strikingly displaying a circular opening [Fig. 3.5(*a*)] and an interior view of its pore space [Fig. 3.5(*b*)] [27].

Returning to the above scheme of the scattering law for a combined mass/pore/surface fractal, the following three equations for exponent f in Eq. (3.10) can be established, designating embedding Euclidian space dimension by E:

$$\text{Property} = \text{mass [Eq. (3.8)]} \qquad \text{or} \qquad f(d, d, E) = -d \tag{3.13a}$$

$$\text{Property} = \text{surface [Eq. (3.9)]} \qquad \text{or} \qquad f(E, d, E) = d - 2E \tag{3.13b}$$

$$\text{Property} = \text{pore} \qquad \text{or} \qquad f(E, E, d) = -d \tag{3.13c}$$

Note that the right-hand side of this array simply gives the exponents of the appropriate scattering power laws of Eq. (3.8) for (purely) mass, of Eq. (3.9) for surface and—additionally—the exponent for scaling pores, as mentioned above.

TABLE 3.1 Property–Distance Scaling Relations of a Purely Mass, Surface, and Pore Fractal [25]

Property	Mass	Surface	Pore	Dimension
Mass	R^d	R^d	R^3	$1 < d < 3$
Surface	R^3	R^d	R^3	$2 \leq d < 3$
Pore	R^3	R^d	R^d	$2 \leq d < 3$

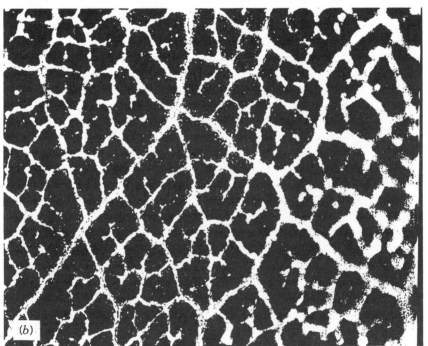

Fig. 3.4 (*a*) Photograph of an area of bare scorched earth. (*b*) Negative of a microphoto-graph of a 2.3×1.3-mm^2 section of a maple leaf.

15 μ

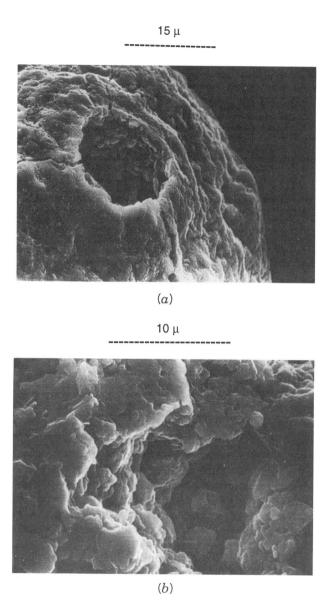

(a)

10 μ

(b)

Fig. 3.5 Scanning electron microscopic images of a spheroidal catalyst particle of 50-μ diameter, showing [(a), ×2000] a circular pore opening of about 15 μ or 1.5×10^5 Å and [(b), ×4000] an interior view of its pore space. [Reprinted with permission from R. Mann and M. C. Wasilewski, *Trans IChemE* **68** A, 177 (1990), Figs. 1, 2, The Institute of Chemical Engineers.]

We compare coefficients a_i of equal powers in d and E between Eq. (3.11) and Eq. (3.13a–c):

$$f(\alpha_1, \alpha_2, \alpha_3) = f(d, d, E) = a_0 + a_1\alpha_1 + a_2\alpha_2 + a_3\alpha_3 = a_0 + (a_1 + a_2)d + a_3E = -d$$

$$f(E, d, E) = a_0 + (a_1 + a_3)E + a_2d = d - 2E \qquad (3.13d)$$

$$f(E, d, d) = a_0 + a_1E + (a_2 + a_3)d = -d$$

As an explanatory example, consider row 1 of Eq. (3.13d): It represents a purely mass fractal, scaling with mass, surface, pore under d, d, E, respectively. (Refer also back to Table 3.1.) These three dimensions now appear in the second and third equalities of row 1 as the arguments α_i of function $f(\alpha_1,\alpha_2,\alpha_3)$ [Eq. (3.11)]. In other words, $\alpha_1 = \alpha_2 = d$, $\alpha_3 = E$. Single term $-d$, in the last equality in row 1, designates the appropriate power in the scattering law for a purely mass fractal [Eq. (3.13a) or (3.8)]. And so on, for the surface and pore fractal in row 2 and row 3, respectively. Equating coefficients a_i in Eq. (3.13d) row by row gives

$$\text{Row 1: } a_1 + a_2 = -1; \; a_0 + a_3E = 0$$
$$\text{Row 2: } a_0 + a_1 + a_3 = -2; \; a_2 = 1 \qquad (3.13e)$$
$$\text{Row 3: } a_2 + a_3 = -1; \; a_0 + a_1E = 0$$

The unique solution of Eq. (3.13e) yields numerical values $a_2 = 1$, $a_1 = -2 = a_3$, $a_0 = 2E$. Consequently, we find [see Eq. (3.11)] $f(\alpha_1,\alpha_2,\alpha_3) \equiv f(d_M, d_S, d_P) = a_0 + a_1 d_M + a_2 d_S + a_3 d_P = 2E - 2d_M + d_S - 2d_P$, yielding the desired scattering intensity wave vector relation for the mixed fractal in the form

$$I(q) \sim q^{d_S - 2(d_M + d_P) + 2E} \qquad (3.14)$$

Checking it out on the example of a purely pore fractal [see Eq. (3.13c)] we obtain for the exponent a value of $d - 2(3 + d) + 6 = -d$, as expected (Table 3.1). Note again the equivalence of fractal dimensions for pore, d_P, and mass, d_M, under $I(q)$: Both dimensions show up under the same coefficient a_i in the exponent of Eq. (3.14).

Surface fractal dimension d_S is restricted to certain limits [25], namely, $d_S \le d_P \le d$, $d_S \le d_M \le d$, where d is the global exponent of the $I(q) - q$ relation of Eq. (3.14). These conditions are readily understood from the nature of the surface as interface between pore and mass for a (purely) pore fractal [see Fig. 3.4(a)] and the reversed case for a mass fractal. An interesting example seems to be the Menger sponge, with $d_S = d_P = d_M = \log 20/\log 3 \approx 2.74$ [28]. Feeding this result into Eq. (3.14) yields scattering intensity wave vector law $I(q) \sim q^{-2.22}$.

3.1.4 A Summary

Small-particle aggregation, forming more often than not a fractal object of considerably more complex, extensive, and useful properties than those exhibited by its constituent units, is raising increasing interest in chemistry. This finding reflects, indeed, a belated concern, recalling that the importance of self-organization of simple entities to form more complex assemblies has been recognized as a fundamental process in biology, physics, astronomy, geology, and so on—not even mentioning its participation in the various social affairs of humanity.

The particular process of diffusion-limited aggregation, DLA, and its extensions play a dominant role [29]. The reason is easy to see: The mechanism, a random search (the DL) leading nevertheless to organized growth patterns (the A), is rather simple to code. In addition, modifications of the details of the generating principle permit a variety of outcomes—such as cluster–cluster aggregation instead of cluster–particle aggregation.

In preceding sections, we have extensively used the mass–distance scaling relation $M(R) \sim R^d$. It is then appropriate to indicate what to do if the functional relation is narrowed to an equality. Hence

$$M(R) \sim R^d \Rightarrow M(R) = A(R)R^d \qquad (3.15)$$

which forces us to determine the nature of prefactor $A(R)$. Consider its following aspects. First, take logarithms in Eq. (3.15), which gives $d = \log[M(R)]/\log(R) - \log[A(R)]/\log(R)$. Clearly, unless quantity $\log[A(R)]/\log(R) \Rightarrow 0$ for $R \Rightarrow 0$, fractal dimension d would depend on the size of the fractal—which cannot be. Second, draw a complete circle on a piece of paper and shade its interior with mass, leaving some spots blank, here and there, to model a mass fractal (in two-dimensional embedding space) [30]. Now, from any point within the disk, draw complete circles of radii $R_1, R_2, \ldots$, around some center, staying within the disk. For each index i, the mass distribution along a circle obeys Eq. (3.15). But note that a circle may fall on unshaded area (empty mass) or several. Hence, $M(R_i)$ is *de*creased or *in*creased with respect to any other $M(R_i)$ if hitting more or less empty mass, respectively. Consequently, $M(R) = A(R)R^d$ *fluctuates* about R^d, we therefore take the root-mean-square deviation V,

$$V[A(R)] = \{\langle [M(R)]^2 \rangle - \langle M(R) \rangle^2\}^{1/2} \langle M(R) \rangle^{-1} \qquad (3.16)$$

where $\langle \cdots \rangle$ indicates an expectation value. Quantity $V[A(R)]$ is an *average* value, called the lacunarity of the fractal, from lacus (Latin) = lake, pool [30].

In light of the above, two recent studies aiming to find an experimental value for prefactor $A(R)$ in mass–distance scaling of cluster–cluster aggregation of aerosols and of soot, respectively, must be read with caution regarding their generalized claims because the proper relation between prefactor $A(R)$ and the lacunarity of a fractal was most likely misread [31, 32].

3.2 FRACTAL ASPECTS OF AGGLOMERATION DYNAMICS

Chemists are concerned with processes, reactions; in summary, with dynamics. They want to know the rate, extent, and degree of product formation during a process. Clearly, then, we must now look at the dynamics of fractal agglomeration phenomena, complementing thereby our previous discussions on fractal aspects of the end product, the static aggregate. No additional theoretical material needs to be included; we are only concerned with dynamics concerning fractal aggregation and fragmentation that take place in ordinary Euclidian spaces, their dimension $E = 2$ or $E = 3$ being solely dictated by the container of the experimental setup, say a Petri dish or a beaker.

Now is a good time to mention the more complicated effects that arise from dynamic phenomena occurring specifically *on* or *within* a fractal structure. In such situations—and depending on the branching/looping pattern of the fractal's structure—classical expressions of transport phenomena such as diffusion, long-wavelength longitudinal modes, and kinetics have to be modified or reinterpreted, leading to the concept of a "fracton" [33]. However, this will not be our concern in this chapter.

3.2.1   Electrolytic Deposition

Electrolytic metal deposition takes place in many technological applications within a three-dimensional fluid volume equipped with three-dimensional electrodes. Furthermore under realistic conditions, convection of electrolyte is present and electrodeposits may grow to appreciable size (several millimeter). Under such scenarios, the aggregation process may not follow the simple mechanisms of cluster aggregation of particle–cluster DLA and cluster–cluster DLA. We recall that they predict aggregate fractal dimension $d \approx 2.5$ for the first, $d \approx 1.7$ for the second in $E = 3$ spaces (Section 3.1.2). Although earlier work (1980) along these lines reported mass fractal dimensions in fine agreement with these predictions, it has become clearer that the simple computer-modeled mechanisms can be no more than approximations of complex, actual agglomeration processes.

Concerning the experimental setup of electrodeposition studies, it is important to describe the form and placement of the electrodes. A small, nearly point-like cathode is favorable for particle–cluster DLA. On the other hand, a strip-shaped cathode facilitates diffusion-limited deposition, with separate clusters growing on each of its many nuclei or seed points. Consequently, the deposition process sprouts a forest (often very pretty) of "fractal trees" [34]. Appropriate model predictions from computer runs accordingly turn out to be considerably more complex than those of DLA on an originally point-like nucleus. For instance, we can evaluate diverse fractal dimensions from mass–distance, mass–time, and perimeter analysis (see Section 3.1.2.1) on a single fractal tree. Contrarily, we may be interested in global measures such as the two-point density–density correlation function between any two locations at different heights of the fractal forest (although we shall not do so).

3.2.1.1   *Silver Aggregation under Damped Electrolyte Convection*
A study on three-dimensional fractal growth of silver electrodeposits forms the basis of our next discussion [35]. The study is of particular interest as effects of sol–gel states (Section 3.1.1) on electrolyte convection near the cathode were examined by adding varying amounts of agarose, a sugar. Furthermore, the electrodepo-

sition runs were followed far beyond initial stages of growth, generating Ag deposits as large as 1 cm. Last, we will see that the reported data permit some ready extrapolations to general principles of fractal growth that reach beyond specific agglomeration dynamics of specific substrates.

The shape of the Pt cathode in the experiments [35] was indeed spherical and small (0.76-mm diameter); as discussed in Section 3.2.1, we therefore expect manifestations of a particle–cluster DLA mechanism. Figure 3.6 shows log–log scaling data of electrodeposition charge Q (~mass M) and linear size r of an Ag aggregate for a gel-like electrolyte solution with agarose as gel agent. The linear least-squares slope corresponds to mass-fractal dimension $d = d_M = 2.53$, rather close to prediction $d_M = 2.5$ of particle–cluster DLA in $E = 3$ (see Section 3.1.2). In addition, the slope exhibits a break at aggregate radius $r = r_C$, signaling a crossover from near-isotropic growth patterns to anisotropic growth of the Ag electrodeposits [35]. Accordingly, r_C relates to the time range of usefully analyzable data; growth data beyond times of electrolysis equivalent to r_C are of questionable value for ready interpretation.

Next, we look at the effects of gel agent agarose. In the agarose-free system and, consequently, minimal damping of electrolyte convection at the cathode, the Ag mass dimension tends to $d_M = 3$. Limit $d_M \approx 3$ is rather remarkable but to be expected because freer convection of electrolyte at the cathode—a mixing effect—facilitates access of the ions to interior regions of the aggregate, the "forbidden" zones during ionic migration under purely DLA. We then note that electrolyte convection

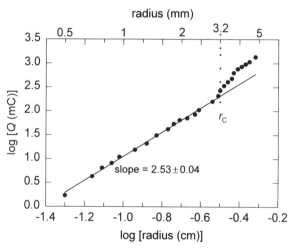

Fig. 3.6 Log–log (decimal) plot of electrodeposition charge (mC) and Ag agglomerate radius over a run of about 1.5 h in an electrolyte of 0.005 M Ag_2SO_4, 0.50 M Na_2SO_4, 0.010 M H_2SO_4, and 0.5 g/100 cm^3 agarose, at constant potential (versus SCE) of -0.200 V. Critical Ag aggregate size r_C is indicated. [Reprinted with permission from P. Carro, S. L. Marchiano, A. Hernández Creus, S. Gonzáles, R. C. Salvarezza, and A. J. Arvia, *Phys. Rev. E* **48**, R2374 (1993), Fig. 3b, The American Physical Society.]

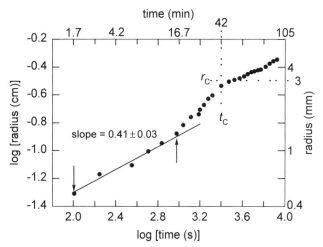

Fig. 3.7 Log–log (decimal) plot of Ag aggregate radius versus electrodeposition time of the gel-like system displayed in Fig. 3.6. The slope and its standard deviation are computed by linear least-squares using data points within range described by the vertical arrows. [Reproduced with permission from P. Carro, S. L. Marchiano, A. Hernández Creus, S. Gonzáles, R. C. Salvarezza, and A. J. Arvia, *Phys. Rev. E* **48**, R2374 (1993), Fig. 1c, The American Physical Society.]

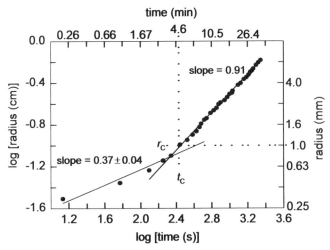

Fig. 3.8 Log–log (decimal) plot of Ag aggregate radius versus electrodeposition time of the agarose-free system. Ionic solution composition as in Figs. 3.6 and 3.7. Critical parameters t_C and r_C are entered. Slopes are computed by linear least-squares from the data points at $t \leq t_C$ and $t \geq t_C$, respectively, with standard deviation for slope $s = 0.37$ given. [Reproduced with permission from P. Carro, S. L. Marchiano, A. Hernández Creus, S. Gonzáles, R. C. Salvarezza, and A. J. Arvia, *Phys. Rev. E* **48**, R2374 (1993), Fig. 1a, The American Physical Society.]

at the cathode partly counteracts a restraining effect exerted by the system on its own growth (see also the beginning of Section 3.1.2). With slowly *increasing* degrees of gel-like character (from *increasing* concentrations of the sugar), mass fractal dimension d_M of the Ag aggregate *decreases* from $d_M = 3$ until its lower limit $d_M = 2.53$ of the gel state is reached at maximum agarose content, as expected.

To get a general idea on overall growth rates of Ag deposition, Figure 3.7 shows the scaling plots of the data for the gel-like, while Fig. 3.8 shows plots for the agarose-free system. Both plots exhibit power law behavior of the form

$$r(t) \sim t^s \tag{3.17}$$

with time exponent s equal to 0.41 ± 0.03 and 0.37 ± 0.04, respectively. There is a crossover time t_C (corresponding to radius r_C) which, expectedly, is shorter for the agarose-free solution because of its larger electrolyte flux.

What can we say about the value of exponent s? How does it relate to the exponents from the mass(charge)–distance scaling shown in Fig. 3.6? In fact, the correspondence is not difficult to establish. We set

$$M(t) \sim t \tag{3.18}$$

because it seems reasonable to assume that the amount of mass accrued during time t is proportional to the number of random steps, $j(t)$, performed by the diffusing ions: $j(t) \sim t$. Then, writing a mass–distance scaling relation [Eq. (3.1)] and relating it to Eq. (3.18), $M(r) \sim r^d \sim M(t) \sim t$, yields

$$r \sim t^{1/d} \tag{3.19}$$

Thus, slope s of the log (radius) − log(time) plot should be equal to the inverse mass fractal dimension $1/d$ [36]. Incidentally, we have met Eq. (3.19) before, albeit in different disguise, namely, as Eq. (2.13). Just consider that during a polymerization reaction—a type of stepping process—the degree of polymerization, given by symbol N in Eq. (2.13), ought to be proportional to run time t.

For the Ag deposition process, this works out as follows: Slopes $s = 0.41$ and 0.37, entered into the radius–time plots of Figs. 3.7 and 3.8, respectively, are close to prediction $s = 1/d$ with $d = 2.53$ (see Fig. 3.6). However, the good agreement pertains only to the range of data points delineated by arrows (Fig. 3.7) or up to time t_C (Fig. 3.8)—about one time decade.

Figure 3.9 offers a remarkable visual aspect of an Ag aggregate electrodeposited in the agarose-free system during run times greatly exceeding t_C—a huge pine cone of mass dimension $d_M \approx 3$ [35].

3.2.1.2 *Formation of Dendrites: Palladium Hydride*

In this section, we are discussing experiments of palladium hydride electrodeposition on a nonpoint cathode [37] under experimental conditions leading to multinuclei or dendritic growth of a fractal forest (see Section 3.2.1).

Fig. 3.9 Photograph of an Ag aggregate grown in the agarose-free system during electrode-position time $t \approx 1$ h. [Reprinted with permission from P. Carro, S. L. Marchiano, A. Hernán-dez Creus, S. Gonzáles, R. C. Salvarezza, and A. J. Arvis, *Phys. Rev. E* **48**, R2374 (1993), Fig. 2a, The American Physical Society.]

Briefly, the cathode was fashioned into a 25-μm thick, 99.9% pure Pd foil. The electrolyte consisted of 50% CH_3OH, 33% HNO_3, and 17% H_3PO_3; the anode was also a Pd foil. (Prior to the actual Pd electrodeposition runs, the cathode was sub-jected to a hydriding process by electrolytic compression of H^+ in a C_2H_5OH/H_2SO_4 medium.) The process was observed by image analysis of the high-resolution trans-mission microscopy data and their fractal analysis performed by convenient cover-ing/counting methods, such as the perimeter scaling approach (see Section 3.1.2.1) or the box counting technique (see Chapter 8).

Figure 3.10 displays four stages of palladium black formed at increasing run times (bottom to top) [37, 38]. Note a tendency of "filling-in," rendering the outline of the forest visually less jagged, an observation quantified by its perimeter (or coastline) fractal dimension decreasing from $d_C = 1.3$ (pattern at bottom) to $d_C = 1.1$ (top). Perimeter dimension $d_C = 1.3$ agrees with $d_C = 1.3$ of silica clusters, as discussed in Section 3.1.2.1, thereby implying that particle aggregation in such diverse systems as precipitated silica and electrolytically deposited PdH neverthe-less obey the same diffusion-controlled mechanism, namely, cluster–cluster DLA. This notion is strengthened by noticing that the observed mass fractal dimension of isolated PdH dendrite structures (bottom pattern), $d_M = 1.80 \pm 0.09$ [37], agrees

Fig. 3.10 Transmission electron microscopic images of Pd hydride electrodeposition on a Pd foil at increasing run times (from bottom to top, time scales unknown), showing gradual smoothing of the dendrite growth patterns under a drop of fractal perimeter dimension from $d = 1.3$ (bottom) to $d = 1.1$ (top). Note the individual dendrites of mass fractal dimension $d = 1.80$, situated about one-third from the left in the bottom pattern. [Reprinted with permission from L. A. Bursill, P. Julin, and F. Xudong, *Internl. J. Mod. Phys.* B **5,** 1377 (1991), Fig. 1, World Scientific Publishing Co.]

with $d = 1.7$–1.8 predicted for cluster–cluster DLA in $E = 3$ space (see Section 3.1.2) [10].

No doubt, electrodeposition processes on foils are far more complex than aggregate formation on and from a small, point-like spherically shaped growth nucleus; it is asking much from computer-generated diffusion-limited aggregation models to take care of all intervening system variables. Nevertheless, the models suggest what to look for and which fractal quantities to measure most profitably. Furthermore, they are indispensable in bracketing possible outcomes of laboratory experiments [39].

We will have an opportunity to return to the surface properties of the electro-deposited PdH aggregates in Section 6.1.2, where we discuss aspects of surface roughness during metal surface corrosion and deposition.

3.2.2   Universal Behavior: Growth of Melanin, Colloidal Gold, and Humic Acid Aggregates

Melanin, a naturally occurring pigment (from Greek mela = black) in everybody's skin, serves as protection against tissue-damaging sun radiation with increasing effi-ciency toward the harmful 3000-Å wavelengths region. Indeed, white skin transmits 25% sunlight, tanned skin only 5% (for the same genotype) [40, 41]. Melanin is synthesized from tyrosine (dihydroxyphenylalanine, or dopa) by the Cu-containing enzyme tyrosinase; Fig. 3.11 gives an idea to the chemist of active groups involved in the overall structure.

But how do fractals come in here? In fact, melanin particles undergo aggregation to large clusters. Furthermore, a mere adjustment of the melanin solution pH en-ables switching between two limiting aggregation processes, namely, DLA and RLA (reaction-limited aggregation), making the substance quite an interesting study object for us.

As the name implies, in reaction-limited aggregation (RLA) the rate-determining step is determined by the reaction rate of cluster–cluster aggregation rather than by the cluster diffusion rate as in DLA. No longer is it assumed—as in the basic DLA model—that clusters instantly and irreversibly stick upon first collision. Hence, mod-ification of DLA algorithms to model many cluster–cluster (or particle–cluster) colli-sions *prior to* irreversible aggregation (see Section 3.1.2) ought to serve rather well as RLA mechanism. However, the method is costly in computer time [14], consider-ing realistic attempt frequencies of potential barrier crossings of the order of 10^{10} s^{-1}. But computational details do not concern us here; we only have to remember the pre-dicted mass fractal dimensions $d_M = 1.53$ ($E = 2$) and $d_M = 1.98$–2.11 ($E = 3$) under the cluster–cluster RLA mechanism. Note that $d_M \approx 2$ ($E = 3$) for RLA is larger than the corresponding $d_M \approx 1.7$ for cluster–cluster DLA ($E = 3$). The reason is simply that species engaged in many collision attempts—running through a series of differ-

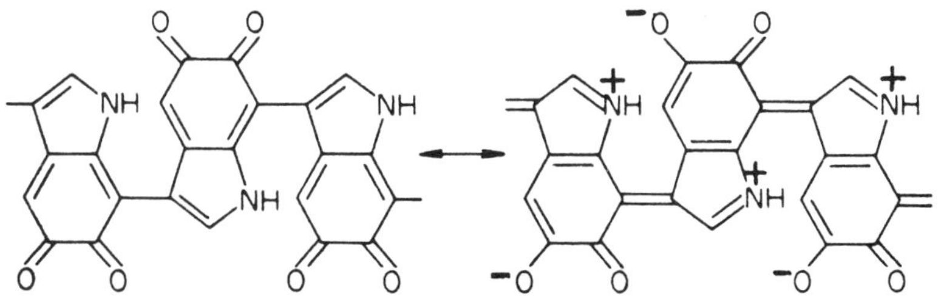

Fig. 3.11   Resonance forms of indolic melanin. [Reprinted with permission from A. E. Needham, *The Significance of Zoochromes* (Springer, New York, 1974), Fig. 3.7 VI, Copy-right 1974 Springer-Verlag.]

ent relative positions—are better able to circumvent steric restrictions from their out-reaching branch structures than the "one-shot" species encounters in DLA.

Figure 3.12 displays radius–time scaling data of a 0.6 wt% aqueous solutions of synthetic, protein-decoupled melanin adjusted to pH 2.96 and 3.4, respectively [41], as obtained by light scattering measurements. The following are the observed, strikingly different fractal aspects depending on the solution pH: (a) pH 3.4: Aggregation of primary clusters of 10–20-Å size follows exponential growth—a distance–time relation we had not encountered before,

$$r \sim \exp(t/\tau) \qquad (3.20)$$

with time constant $\tau = 2.96$ h (time interval to increase the momentary cluster radius r by a factor 2) and mass fractal dimension $d_M = 2.2$ [Eq. (3.8)]. This slow exponential growth as well as mass fractal dimension $d_M = 2.2$ of the larger aggregates imply re-action-limited aggregation (RLA)—multiple cluster–cluster collision attempts prior to an irreversible aggregation step—under its predicted $d_M = 1.98$–2.11 (see above). (b) pH 2.96: The aggregation process has switched to the rapid power law growth with slope $s = 0.56 \pm 0.01$, yielding aggregate mass fractal dimension $d_M = 1/s = 1.80 \pm 0.05$ [Eqs. (3.17) and (3.19)]. In turn, this value is quite close to $d_M = 1.7$–1.8 predicted for cluster–cluster DLA in space $E = 3$ (see Section 3.1.2). To repeat then, change of the hydrogen ion concentration by a mere factor of 3 switches the aggregation mechanism of melanin between diffusion-limited and reaction-limited mechanisms, with the RLA mechanism prevailing at the lower H^+ concentration.

The reader who may think that the transition between the DLA and RLA process of melanin aggregation is a rather specialized and unique phenomenon, will be

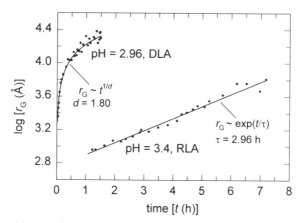

Fig. 3.12 Semi-log (decimal) plot of the radial growth rate of melanin clusters by diffusion-limited and reaction-limited cluster–cluster aggregation. [Reprinted with permission from J. S. Huang, J. Sung, M. Eisner, S. C. Moss, and J. Gallas, *J. Chem. Phys.* **90**, 25 (1989), Figs. 2, 3, Copyright 1989 American Institute of Physics.]

surprised to learn that the results do not differ by one iota from those recorded during aggregation of aqueous colloidal suspensions of gold, readily steered between RLA and DLA by changes in the concentration of added pyridine [42]. No need to describe the results, to list the mass fractal dimensions or power exponents, to draw plots: Just replace 0.6 wt% aqueous melanin by 5000 aqueous-colloidal primary Au particles and substitute pH 3.4 by 10^{-5} M pyridine (RLA) and pH 2.96 by 10^{-2} M pyridine (DLA). Everything else stays essentially the same, even time constants τ of reaction-limited growth [Eq. (3.20)] are equal to each other within 25%.

Figure 3.13 exhibits transmission electron microscopic images of the Au aggregates for DLA (*a*) and RLA (*b*), giving a fine visualization on the strikingly different aggregate densities and structures under the two cluster–cluster aggregation schemes.

Well, what to make of this? Certainly, it shows a commonality and unifying general aspect of fractality that reaches over objects bearing large structural-chemical differences. It is quite useful to have a predictive number of an aggregation scheme—its mass fractal dimension—that so readily permits cataloguing and correlation among different objects [43]. On the other hand, a universality implies ignorance of the detailed, system-related mechanisms: Does this signify that the molecular interactions leading to aggregation of suspensions of colloidal gold in aqueous pyridine on one hand, and of synthetic melanin in acidified water on the other, are identical? Hardly. What this agreement means is that the overall kinetics of the aggregation processes of the two different systems are essentially *indistinguishable on the resolution level of the applied experimental observations.*

Finally, we briefly consider results [44] of aggregation phenomena of water-

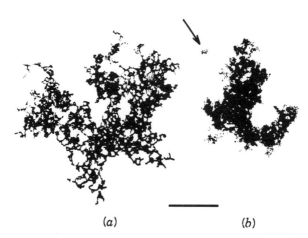

(*a*)   (*b*)

Fig. 3.13 Transmission electron microscopic images of clusters of 5000 primary Au colloidal particles suspended in water by diffusion-limited (*a*, 10^{-2} M pyridine) and reaction-limited (*b*, 10^{-3} M pyridine) aggregation. The horizontal delineates 5000 Å, the arrow points to a cluster remnant that has not joined the global aggregate. [Reprinted with permission from D. A. Weitz, J. S. Huang, M. Y. Lin, and J. Sung, *Phys. Rev. Lett.* **54,** 1416 (1985), Fig. 2, American Physical Society.]

suspended humic acid, a heterogeneous mixture of organic polyelectrolytes and an important component of soils. This system tunes crossover from DLA to RLA aggregation by means of the *temperature,* as indicated by its mass fractal dimensions $d_M(DLA) = 1.8$ (4 °C) and $d_M(RLA) = 2.3$ (22 °C); the DLA $\Rightarrow$ RLA switch-over having reached saturation after 34 h.

Hence, we have now discussed three agents that cause crossover between diffusion-limited (DLA) and reaction-limited aggregation (RLA): (a) Solution pH in the aggregation of melanin in water, (b) concentration of pyridine in the aggregation of aqueous suspension of colloidal gold, and (c) ambient temperature in the aggregation of humic acid.

3.2.3 A Summary

Agglomeration or clustering of smaller particles to form a larger object is one of the moving processes leading to fractal structures of interest and relevance to chemistry. Although the mechanisms leading to the global aggregate(s) consist of random-diffusional events performed by smaller entities, the final structure exhibits a degree of self-organization.

To exemplify this, first recall the diffusion-limited particle–cluster aggregation process. We learned that the outreaching branches of the growing aggregate effectively close its more centrally located regions to oncoming random walkers: The uniformly available Euclidian space is filled under some *restriction* that forbids its complete filling. Yet, filling is evidently optimal within this constraint, as the dimension of the cluster exceeds that of its constituent pieces.

Second, recall aggregation of silver by electrodeposition of its sulfate (Section 3.2.1.1). The situation is reversed here: Once restrictions by the cathode cluster geometry on diffusive aggregation are lifted by increased convective transfer, the random process is seen to fill available Euclidian space fully (mass fractal dimension $d_M \approx 3$).

Hence, I spell out what may serve as a comprising description of fractal phenomena discussed so far:

> A fractal structure often results from a random, iterative, but optimally space-filling growth process under some constraint.

Clearly, the resulting fractal dimension must be inferior to that of its immediately embedding Euclidian space and a point can be made that self-similarity (statistical) of such growth structure as well as lower and upper limits of the scale over which the process is fractal—ideally from the average size of the constituent parts to the radius of gyration—are then system-given consequences. Further, I think that this definition goes beyond that proposed on tenets of randomness [43]; a *constraint* on the random growth process is needed for formation of *fractal* structures. Otherwise, the embedding Euclidian space is filled.

Incidentally, the effects of control exerted by fractal growth on its further progress has thought-provoking parallels in the (lower) animate world: A colony of bacteria growing on agar plates in a diffusion-limited nutrient supply assumes a statistically self-similar structure as the bacteria in the most advanced regions 'appropriate' the food resources; on the other hand, if the nutrient supply is freely accessible, the front of the advancing colony assumes a self-affine shape reflecting random walk under a positive bias [45].

NOTES AND REFERENCES

[1] We may use the Koch surface (Section 2.2) as a convenient model of a surface fractal. As a simple model of a pore fractal, take a solid with a cavity in the pattern of a Koch curve.

[2] A long time ago, I worked at a place called Electronized Chemicals Corporation (Brooklyn, NY), one of the commercial laboratories to benefit from venture capital (shortly after WW II). The company developed a method for sterilization of food stuff and other perishable materials by ionizing radiation for purposes of subsequent shelf storage without refrigeration. The radiation consisted of ultra-short pulses of electron beams of about 1 MeV and thousands of ampère, penetrating about 1 cm of a water layer. (I measured the dosage by recording the temperature rise of an irradiated water volume contained in a Dewar with a lab thermometer, also serving as manual stirrer! No, no, *after* the final electron pulse.) Obviously, it offered fascinating frontier research: One day, someone—from the Brooklyn Polytechnic Institute (as it was then called)—brought samples of a monomer in thin glass tubes (I forget which, but its polymer is soluble in the monomer), merely sealed under nitrogen but not having received other special treatment, as best as I can recall. After receiving a certain radiation dose (the glass was slightly browned by F-centers) and after *several months* of leaving the tubes undisturbed in a dark place, an accelerating increase in viscosity indicated a polymerization reaction—thereby hinting at free radicals *of lifetimes of the order of months* in an initially nonviscous liquid *and* at room temperature.

[3] M. Daoud and J. E. Martin, *Fractal Properties of Polymers,* in *The Fractal Approach to Heterogeneous Chemistry,* D. Avnir, Ed. (Wiley, Chichester, UK, 1989), p. 109.

[4] R. Eksteen and K. J. Pardue, *Modified silica-based packing materials for size exclusion chromatography,* in *Handbook of size-exclusion chromatography,* C. Wu, Ed. (Dekker, New York, 1995), p. 47.

[5] M. le Maire, A. Ghazi, M. Martin, and F. Brochard, Calibration curves for size exclusion chromatography: Description of HPLC gels in terms of porous fractals, *J Biochem.* **106,** 814 (1989).

[6] For instance, see Eq. (2.10), replacing film thickness z by length R. Alternately, integrate Eq. (2.7), replacing pore radii ρ by R. Equation (3.3) may also be looked upon as all fractal surface cross-sectional elements R^{-d+2} summed between R and L and normalized to L, $K_d = \int_R^L R^{2-d}\, dR / \int_0^L L^{2-d}\, dL$.

[7] F. Brochard, A. Ghazi, M. le Maire, and M. Martin, Size exclusion chromatography on porous fractals, *Chromatographia* **27,** 257 (1989). Solvents induce swelling of polymer chains: In a good solvent ($\upsilon = \frac{3}{5}$), interlink repulsive forces prevail, in a poor solvent ($\upsilon = \frac{1}{3}$), interlink attractive forces prevail, and a theta solvent ($\upsilon = \frac{1}{2}$) manifests their balance.

[8] $M_L = \sup M(R = L)$; see *covering,* Glossary 1.

[9] I. Porcar, R. García, V. Soria, A. Campos, and J. E. Figueruelo, Porous fractal gels: secondary effects in SEC, *J. Non-Cryst. Sol.* **147, 148,** 170 (1992).

[10] T. A. Witten, Jr., and L. M. Sander, Diffusion-limited aggregation, a kinetic critical phenomenon, *Phys. Rev. Lett.* **47,** 1400 (1981); T. A. Witten and L. M. Sander, Diffusion-limited aggregation, *Phys. Rev. B* **27,** 5686 (1983); for some dazzling computer images, see T. Wegner and M. Peterson, *Fractal Creations* (Waite Group Press Inc., Coste Madera, CA, 1991), p. 178.

[11] H. Takayasu, *Fractals in the physical sciences* (Manchester University Press, Manchester, UK, 1990), p. 61.

[12] P. Meakin and Z. R. Wasserman, Fractal structure of three-dimensional Witten-Sander clusters, *Chem. Phys.* **91,** 391 (1984).

[13] There is more to report here on the past doings at Electronized Chemicals Corporation (see [2] above). We irradiated sheets of polymethyl methacrylate with a small electron beam dose that generated an essentially two-dimensional, now so familiar DLA-like pattern of dielectric breakdown at a depth inside the polymer sheet corresponding to maximum energy dissipation (about 3-mm depth from the surface, if I remember correctly), called Lichtenberg figure. Of course, we could not have known then (*1950*) that we looked at a real *fractal* pattern. [See L. Niemeyer, L. Pietronero, and H. J. Wiesmann, Fractal dimension of dielectric breakdown, *Phys. Rev. Lett.* **52,** 1033 (1984).] Trimmed and mounted vertically over a fluorescent bulb, these things made beautiful and impressive gifts for visiting VIP. I never got one, so I cannot impress this readership by a photograph of such precocious exposure to fractal phenomena.

[14] P. Meakin, *Simulation of aggregation processes,* in *The Fractal Approach to Heterogeneous Chemistry,* D. Avnir, Ed. (Wiley, Chichester, UK, 1989), p. 131.

[15] The largest clusters generated by this process are no longer fractals, a situation not unlike that causing crossover from fractal to Euclidian dimensions in particle–cluster DLA (Section 3.1.2).

[16] A. Erzan, L. Pietronero, and A. Vespignani, The fixed-scale transformation approach to fractal growth, *Rev. Mod. Phys.* **67,** 545 (1995). The leading seven-to-eight pages offer a good descriptive introduction and summary of the physics of fractals and the properties of fractal growth models.

[17] D. W. Schaefer, What factors control the structure of silica aerogels?, *Rev. Phys. Appl.* **24,** C4-121 (1989).

[18] A. Tuel, P. Dautry, H. Hommel, A. P. Legrand, and J. C. Morawski, Image analysis of fractal aggregates of precipitated silica, *Prog. Coll. Polym. Sci.* **76,** 32 (1988). There is a discrepancy between the author's value of $d \approx 1.8$ and that derived from the least-squares calculation, $d = 1.5$, from Fig. 4 of [18].

[19] B. H. Kaye, *Image analysis techniques for characterizing fractal structures,* in *The Fractal Approach to Heterogenous Chemistry,* D. Avnir, Ed. (Wiley, Chichester, UK, 1989), p. 55.

[20] *"How long is the coast of Britain?"* (see p. 25 of [2] in Section 1.1). The measuring procedure is also called step divider method; see A. M. Hammad and M. A. Issa, Fractal dimension as a measure of roughness of concrete fracture trajectories, *Adv. Chem. Bas. Mat.* **2,** 169 (1994).

[21] R. W. Detenbeck, *Light scattering,* in *Encyclopedia of Physics,* R. G. Lerner and G. L. Trigg, Eds. (VCH Publishers, New York, 1991), p. 635.

[22] P. A. Bonczyk and R. J. Hall, Fractal properties of soot agglomerates, *Langmuir* **7,**

1274 (1991). The range of experimental values of wave vector q, $3 < q < 25$ μ^{-1}, is equivalent to a resolution range of 0.04–0.33 μ, a factor of 8. A recent light scattering study [A. A. Vertegel, S. V. Kalinin, N. N. Oleynikov, and Yu. D. Tret'yakov, The fractal particles of iron (III) hydroxonitrate: from solution to solid state, *J. Non-Cryst. Sol.* **181,** 146 (1995)] applies covering range $0.00114 < 1/\lambda < 0.00130$ through a sweep of wavelength λ (rather than scattering angle θ). Such a narrow interval is not sufficient to draw convincing experimental conclusions about fractal aspects of the system.

[23] S. K. Sinha, Scattering from fractal structures, *Physica D* **38,** 310 (1989).

[24] I. Snook and P. McMahon, Fractal pore surfaces in brown coal, their changes on processing and their effect on combustion, *Langmuir* **9,** 2726 (1993). Because wavelengths of incident X-radiation are considerably shorter than those of optical laser-light sources, scattering angles θ (reciprocal wave vectors q^{-1}) are correspondingly smaller (larger) [see Eq. 3.6)].

[25] P. Pfeifer, *Small-angle scattering laws for intermediates between surface, mass, and pore fractals,* in *Multiple Scattering of Waves in Random Media and Random Rough Surfaces* (Pennsylvania State University, University Park, PA 1985), p. 45. Note a typo in Eq. (2): It should read "for *surface* fractal."

[26] See also pp. 30–31 of [12] in Section 1.2.

[27] R. Mann and M. C. Wasilewski, Towards a fractal computer graphic basis for characterisation of catalyst pore structure by image reconstruction, *Trans IChemE* **68** A, 177 (1990). I am grateful to Dr. R. Mann for permission to reprint the photographs.

[28] See p. 145 of [2], cited in Section 1.1.

[29] Some unsuspected details of DLA have turned up upon modeling of increasingly large clusters; see H. Kaufman, A. Vespignani, B. B. Mandelbrot, and L. Woog, Parallel diffusion-limited aggregation, *Phys. Rev. E* **52,** 5602 (1995).

[30] J.-F. Gouyet, *Physiques et structures fractales* (Masson, Paris, 1992), p. 25. The corresponding discussion in Section 34 in the book of Mandelbrot (see [2], Section 1.1)—who coined and defined the term—is in my opinion less accessible.

[31] M. K. Wu and S. K. Friedlander, Note on the power law equation for fractal-like aerosol agglomerates, *J. Coll. Interface Sci.* **159,** 246 (1993).

[32] Ü. Ö. Köylü, Y. Xing, and D. E. Rosner, Fractal morphology analysis of combustion-generated aggregates using angular light scattering and electron microscope images, *Langmuir* **11,** 4848 (1995). This paper gives an impressive demonstration of the difficulties, uncertainties, and complexities of measuring "fractal" prefactor $A(R)$. However, the purported equal footing of prefactor and fractal dimension within fractal theory and praxis misses a very important, universal point of fractality—namely, its scale independence. On the other hand, it is certainly technologically important to know the prefactor of fractal aggregates if the analysis, specifically, strives to establish radii of gyration. It is not certain if the authors had fully appreciated the difference between prefactor and lacunarity.

[33] S. Alexander and R. Orbach, Density of states on fractals: ≪fractons≫, *J. Phys. Lett.* **43,** L-625 (1982).

[34] See Figs. 7a and b of [14].

[35] P. Carro, S. L. Marchiano, A. Hernández Creus, S. Gonzáles, R. C. Salvarezza, and A. J. Arvia, Growth of three-dimensional silver fractal electrodeposits under damped free convection, *Phys. Rev. E* **48,** R2374 (1993).

[36] J. E. Martin, J. P. Wilcoxon, D. Schaefer, and J. Odinek, Fast aggregation of colloidal

silica, *Phys. Rev. A* **41,** 4379 (1990); D. Asnaghi, M. Carpineti, M. Giglio, and M. Sozzi, Coagulation kinetics and aggregate morphology in the intermediate regimes between diffusion-limited and reaction-limited cluster aggregation, *Phys. Rev. A* **45,** 1018 (1992).

[37] L. A. Bursill, P. Julin, and F. Xudong, Fractal analysis of electrolytically-deposited palladium hydride dendrites, *Internl. J. Mod. Phys. B* **5,** 1377 (1991). See also P. Julin and L. A. Bursill, Dendritic surface morphology of palladium hydride produced by electrolytic deposition, *J. Sol. State Chem.* **93,** 403 (1991), a complementary paper dealing with more purely electrochemical aspects of this work. It is the first report under [37] that holds our main interest.

[38] Actual electrodeposition times are not reported in Fig. 1 of [37].

[39] The reader particularly interested in this field may want to look at a model of "random successive growth," which dispenses with the long-range diffusion assumption of DLA and is controlled by the variations of *local* conditions of (a) the growth probability of neighboring sites around the cluster, (b) the occupation ratio of cluster sites within a (small) circle centered on a potential growth site: W. Ziqin and L. Boquan, Random successive growth model for pattern formation, *Phys. Rev. E* **51,** R16 (1995).

[40] A. E. Needham, *The Significance of Zoochromes* (Springer, New York, 1974), p. 59.

[41] J. S. Huang, J. Sung, M. Eisner, S. C. Moss, and J. Gallas, The fractal structure and the dynamics of aggregation of synthetic melanin in low pH aqueous solutions, *J. Chem. Phys.* **90,** 25 (1989).

[42] D. A. Weitz, J. S. Huang, M. Y. Lin, and J. Sung, Limits of the fractal dimension for irreversible kinetic aggregation of gold colloids, *Phys. Rev. Lett.* **54,** 1416 (1985).

[43] D. Avnir, O. Biham, D. Lidar (Hamburger), and O. Malcai, *On the Abundance of Fractals,* in *Fractal Frontiers,* M. M. Novak and T. G. Dewey, Eds. (World Scientific, New York, 1997), p. 199; D. Hamburger, O. Biham, and D. Avnir, Apparent fractality emerging from models of random distributions, *Phys. Rev. E* **53,** 3342 (1996).

[44] R. Österberg, L. Szajdak, and K. Mortensen, Temperature-dependent restructuring of fractal humic acids: A proton-dependent process, *Environm. Interntl.* **20,** 77 (1994). The fractal dimensions were determined by SANS (small-angle neutron scattering); see Eq. (3.8). For possible interference by polydispersity, see S. Z. Ren, E. Tombácz, and J. A. Rice, Dynamic light scattering from power-law polydisperse fractals: Application of dynamic scaling to humic acid, *Phys. Rev. E* **53,** 2980 (1995).

[45] T. Vicsek, M. Cserzö, and V. K. Horváth, Self-affine growth of bacterial colonies, *Physica A* **167,** 315 (1990). Biased random walk will occupy us extensively in later sections.

4

DIFFUSION AND REACTIONS

4.1 INTRODUCTION: THE FRACTON

In this chapter we discuss principles and experimental results of transport phenomena and related properties *on* fractals, such as diffusion, spectroscopy, excitation transfer, reaction rates, and chemical reactivity. How will the well-known descriptions and theories of these processes have to be modified once an underlying substrate is no longer a Euclidian space (as it was in Chapters 1–3) but a fractal? Here, it is assuring to note that we will only concentrate on one basic theoretical example among these transport mechanisms as they all relate to an ancestral differential equation $(1/\zeta)\partial X/\partial t = \partial^2/\partial x^2 + \partial^2/\partial y^2 + \partial^2/\partial z^2$, where X stands for temperature or mass of chemical reagent or $\cdots$ and ζ is the appropriate thermometric conduction κ (Maxwell) or diffusion constant K_0 (Einstein) or ..., in units of (length)^2/time. We begin with diffusion, using the principles of _random walk_. Thereafter, any pertinent flux mechanism on fractal substrates can be taken up with little additional conceptional difficulties.

Keep in mind how the controlling agent has changed: It is no longer the random walker that "distills" a fractal from its Euclidian walking space (such as in DLA); now it is the fractal pattern that *controls* the particulars of the random walker traversing it. Differently put, in the first situation an *object* is created, in the second instance *numerical information* about its structural pattern is obtained. In principle, this implies a significant informative-fundamental step ahead because a fractal dimension does not predict the details of the fractal's structure.

It is convenient to introduce some essentials of the theory by a thought experiment using familiar chemistry [1]. Imagine a very long, coiled linear polymer chain and induce a local excitation, say a long-lived charge pulse, at some point $\mathbf{r}_0$ on it. Now imagine the random progress of the slowly decaying pulse along the polymer chain by clocking time $t = \tau$ once it has exited an area $\pi\xi^2$ determined by (large) root-mean-squared diffusion distance ξ counted from $\mathbf{r}_0$.

Next, consider a second polymer that is sufficiently (but not excessively) cross-linked to contain a collection of statistically self-similar fractal loops, branches, and sets of dangling bonds. Repeating the charge induction experiment on this substrate, should we expect that the two charge pulses exit same area $\pi\xi^2$ at the same time τ?

Hardly. The randomly diffusing charge on the cross-linked polymer chain is likely, during some time intervals of its walk, to run around over loop structures and to get lost at the ends of branches; no such constraints hinder a random walker on the linear polymer. In other words, the random walker on the looped/branched substrate is likely to arrive *later*.

To quantify this, we take Einstein's law of diffusion in Euclidian spaces [2], $\xi^2 \sim \langle R^2 \rangle \sim K_0 t$, and modify it to account for a possibly delayed progress of a random walk on some appropriate fractal, as just described. Formally, we write

$$\xi^2 = \langle R^2 \rangle \sim K_0 t^{2/D_w} \tag{4.1}$$

with condition $D_\mathrm{w} > 2$ (*anomalous diffusion*) [3]. Compared to normal or regular diffusion, exponent value $2/D_\mathrm{w} < 1$ would indeed require a longer τ for exiting equal circular area $\pi\xi^2$ [4]. Conveniently, we change the inequality into an equality by setting $D_\mathrm{w} = 2 + \eta$, $\eta > 0$. Then, Eq. (4.1) reads [5]

$$\xi^2 \sim \langle R^2 \rangle \sim K_0 t^{2/(2+\eta)} \approx K_0 t^{1-\eta/2} \tag{4.2}$$

to first order in $\eta < 2$, which effectively transforms the diffusion *constant* K_0 into the *time-dependent* diffusion *coefficient* [6]

$$K_0(1 - \eta/2)t^{-\eta/2} \approx K(t) \sim t^{-\beta} \tag{4.3}$$

setting $\eta/2 = \beta$, with $0 < \beta < 1$. Therefore, diffusional delay during Brownian motion on substrates forcing the walker into (many) detours is equivalent to a decreasing diffusion coefficient under lengthening observation. [A random walker on a Koch curve (see Fig. 2.6) would *not* experience diffusional delay as there are no detours, merely self-similar "stairs," to negotiate.]

To finish up, we relate one of the two equivalent parameters, D_w or η, to dimension d of the fractal. We do this by some elementary scaling considerations—not much different in principle from what we did in preceding sections. We scale two volume–distance relations, one based on the volume spanned by the random walk of a particle on the fractal of dimension d, the other based on volume–distance scaling of the fractal itself. The first process necessarily contains a new parameter, designated by $\bar{d}$, because the walk now traverses the fractal (a subspace of Euclidian space). The other scaling process involves dimensions d and D_w. Consequently, a comparison of the two outcomes ought to yield a relation between d, D_w, and $\bar{d}$.

First, to scale the volume–distance relation of the random walk on the fractal, we make use of the probability $p(t,\mathbf{r}_0)$ of the random walker to return to its starting point (arbitrarily set $\mathbf{r} = \mathbf{r}_0 = 0$) after time interval t, $p(t,0) \sim t^{-E/2}$ [7] (see also Fig. 1.2). To modify it for walk on fractals, we replace Euclidian dimension E by parameter $\bar{d}$ [3], as mentioned above. Obviously, the smaller $p(t,0)$, the wider the random walker's excursions. Hence, we consider the reciprocal of $p(t,0)$, $[p(t,0)]^{-1}$, as *volumetric measure* taken during a random walk on a fractal:

$$[p(t,0)]^{-1} \sim t^{\bar{d}/2} \tag{4.4}$$

Second, to scale volume $V(\xi)$ of a fractal of dimension d with distance ξ, we use general mass–distance relation Eq. (2.2), $V(\xi) \sim \xi^d$. Then, relating the results of the two methods, we get the sequence of relations

$$[p(t,0)]^{-1} \sim t^{\bar{d}/2} \sim (\langle R^2 \rangle^{1/2})^d \sim (t^{1/D_w})^d \tag{4.5}$$

The first two terms represent the volume–time scaling of the random walk *on the fractal* [Eq. (4.4)], the third and fourth arise from the volume–distance scaling $V \sim \xi^d \equiv [\langle R^2 \rangle^{1/2}]^d$ *of the fractal* and from Eq. (4.1). Comparing time exponents in Eq. (4.5) yields the sought-after relation between $\bar{d}$, d, and D_w:

$$\bar{d} = 2d/D_w = 2d/(2 + \eta) \tag{4.6}$$

Dimension $\bar{d}$ is called fracton or *spectral dimension* of a fractal of dimension d [8]. Parameter D_w has been called the fractal dimension of the path of the random walk [3] although this designation is not helpful here; term η in excess of $D_w = 2$ (see above) seems more apt to a descriptive notation for anomalous diffusion [4].

As $d \leq E$ and $D_w \geq 2$, we obtain the inequality

$$\bar{d} \leq d \leq E \tag{4.7}$$

In principle, $\bar{d}$ may be smaller or larger than 2 [9]: The literature designates case $\bar{d} < 2$ (or $E < 2$) as *compact* or *recurrent* walk, $\bar{d} > 2$ (or $E > 2$) as *noncompact* or *nonrecurrent* walk, with $\bar{d} = 2$ (or $E = 2$) designating the crossover between both diffusion regimes.

For the curious, the fracton dimension of the Sierpinski gasket embedded in $E = 2$, a frequently used prototype of a deterministic looped fractal of dimension $d = \log 3/\log 2 \approx 1.58$, is $\bar{d} = \log 9/\log 5 \approx 1.37$, $D_w = \log 5/\log 2 \approx 2.32$, thus $\eta \approx 0.32$. For percolation clusters embedded in $E = 2$ or $E = 3$, fracton dimension $\bar{d} \approx 1.3$ [3].

4.2   FRACTALITY IN DYNAMICALLY DISORDERED SYSTEMS

We now consider transport processes in fractal systems, making best use of the available predictions of fracton behavior from computer models, scaling arguments, and theoretical enumeration methods. Although computer and scaling models are conveniently based on deterministic and ordered fractals, whereas real systems are irregular, the approach is nevertheless very effective: Recall that fracton behavior depends on the pattern of the fractal's structure, its *connectivity* [10] arising from branching, looping, and so on (see Section 4.1); in other words, the nearest-neighbor relations in reference to a given point on the connected fractal, such as *two* for a point on a segment, *three* or more for a branching point, *one* for the end-point of a dangling bond. Hence, we can deform a model fractal into greater like-ness to a realistic substrate but still accept fracton dimension $\bar{d}$ from the model if the basic makeup of the connectivity of the fractal substrate is retained [11].

Concepts of (self-similar) patterns of loops, branches, and so on, represent a lo-cal connectivity not restricted to fractal substrates but also arising in non-fractal, in-homogeneous substances. The strict definition of a fracton as descriptor of the con-nectivity of a *fractal* object may then be diluted, but the widened concept is very useful. Indeed, we are reminded here of the extension of a mass (M)–distance(R) scaling relation $M(R) \sim R^d$ (under fractal dimension d) to a property(P)–scale scal-ing relation $P \sim \text{scale}^\beta$ (under power law exponent β) of materials not necessarily fractal (see, e.g., Section 2.4.1). [No confusion should arise between this β and that of Eq. (4.3) as they are well defined within their contexts.]

We therefore study transport phenomena in chemical systems by first looking for any anomalous behavior; thereafter we worry whether the dimension of the process represents a fracton—noting well that a fracton dimension $\bar{d} < 2$ predicts anom-alous transport but anomalous transport does not necessarily predict a fracton [12].

4.2.1 Diffusion-Limited Reaction Systems A + A $\Rightarrow$ Product

We had extensively discussed diffusion-limited processes occurring within Euclid-ian spaces and now turn our attention to anomalous diffusion-limited reactions, that is, chemical reactions where diffusion of reactants toward each other again fur-nishes the rate-determining step but no longer follows the classical diffusion equa-tion. First, let us discuss the simple binary reaction scheme A + A $\Rightarrow$ 0, the 0 indi-cating that there is no significant back-reaction rate and that the nature of the reaction product is of no interest. Reaction A + A $\Rightarrow$ 0 seems a rather simple model to use, but the outcome is quite surprising.

The necessary modifications of the classical rate expression $-d[\text{A}]/dt = K_0[\text{A}]^2$ for a fractal medium are readily applied with the help of the results of Section 4.1; we simply change rate constant K_0 into time-dependent diffusion coefficient $K(t) \sim t^{-\beta}$ [Eq. (4.3)]. Conveniently, we express concentration [A] by density $\rho(t)$ (volume^{-1}) and thus obtain rate expression

$$-d\rho/dt = K_0 t^{-\beta} \rho^2 \quad 0 \le \beta < 1 \tag{4.8}$$

for bimolecular reaction [A] + [A] $\Rightarrow$ 0 in a fractal medium [13, 14]. Integrating Eq. (4.8) yields [15]

$$\rho^{-1} - \rho_0^{-1} = (1 - \beta)^{-1} K_0 t^{1-\beta} \tag{4.9}$$

with $\rho(t = 0) \equiv \rho_0$. Equation (4.9) compares to $\rho^{-1} - \rho_0^{-1} = K_0 t$ for the classical system in Euclidian space ($\beta = 0$).

A chemist may find it more convenient writing rate expression $-d\rho/dt$ solely in terms of particle density $\rho(t)$. To do this, (a) we solve Eq. (4.9) for t and (b) replace t in Eq. (4.8) by the resulting expression for t. Step (a) yields $t = [(1 - \beta)/K_0]^{1/(1 - \beta)} \rho^{-1/(1 - \beta)}$; the corresponding term for ρ_0^{-1} is negligible due to condition $\rho^{-1} \gg \rho_0^{-1}$ at longer times. Step (b) yields

$$-d\rho/dt \sim \rho^{2 + \beta/(1 - \beta)} \sim \rho^{1 + 1/(1 - \beta)} \tag{4.10a}$$

Note that if $\beta = 0$ (regular diffusion) we regain classical rate expression $-d[\mathrm{A}]/dt \sim [\mathrm{A}]^2$. On the other hand, for anomalous diffusion, $0 \leq \beta < 1$, term $1/(1 - \beta)$ exceeds unity and may become rather large for $\beta \Rightarrow 1$. Hence, the reaction order exceeds 2, perhaps greatly, although the reaction *is* bimolecular [14]: Presumably, many a chemist doing such type of experiment and being unaware of the peculiarities of anomalous diffusion effects may give up and throw the data into the waste basket.

Suppose reaction $\mathrm{A} + \mathrm{A} \Rightarrow 0$ proceeds on a fractal of fracton dimension $\bar{d} \approx$ 1.4, a percolation cluster [8, 16]. Then, exponent β of Eq. (4.8) equals $1 - \bar{d}/2$, as is readily ascertained by relating volume $\rho^{-1} \sim t^{1 - \beta}$ [Eq. (4.9)] with exploration volume $[p(t,0)]^{-1} \sim t^{\bar{d}/2}$ [Eq. (4.5)]:

$$t^{1-\beta} \sim t^{\bar{d}/2} \qquad 1 - \beta = \bar{d}/2 \tag{4.10b}$$

Hence, Eq. (4.10a) is modified to

$$-d\rho/dt \sim \rho^{1 + 2/\bar{d}} \sim t^{(2 + \bar{d})/\bar{d}} \approx \rho^{2.43} \tag{4.10c}$$

The experimentally found reaction order amounts to 2.43 although the mechanism is, by construction, strictly bimolecular without back-reaction or inhibiting steps.

4.2.2 Diffusion-Limited Reaction Systems A + B ⟹ Product

To explore the behavior of reaction $\mathrm{A} + \mathrm{B} \Rightarrow 0$ on a fractal substrate, it is necessary to recall its standard expression, $-d\rho_A(t)/dt = K_0 \rho_A(t)\rho_B(t)$, and its solutions in the mean-field approximation (absence of spatial fluctuations of species A, B and assuming the long-time regime), namely, $\rho_A(t) \sim K_0 t^{-1}$ for condition $\rho_A(t = 0) \equiv \rho_A(0) = \rho_B(0)$ and $\rho_A(t) \sim \exp\{-K_0[\rho_B(0) - \rho_A(0)]t\}$ if $\rho_A(0) < \rho_B(0)$.

For diffusion-controlled reactions $\mathrm{A} + \mathrm{B} \Rightarrow 0$, more complex behavior appears than for reaction $\mathrm{A} + \mathrm{A} \Rightarrow 0$—regardless of the fractality of the diffuser's space. First, the decay (diffusing-away) of randomly fluctuating interspecies concentration differences $\rho_A - \rho_B$ may become rate determining. Second, local cluster formation within each species A, B no longer guarantees random, uniformly distributed reactive encounter probabilities between the diffusing A and B; therefore a *fractal* dimension may now enter into the rate expression [17–19]. To understand the effect of random reactant density fluctuations [18], which show up toward depletion of reactants (at long reaction times), refer to Fig. 4.1. It depicts the frequently occurring condition $\rho_A(0) = \rho_B(0)$ (batch process), e.g., by 12 concentration units each of molecules A and B, homogeneously and randomly distributed on the *macroscopic* level. On a *local* level, however, the random fluctuations of molecular positions cause imbalances from A = B occupations, sketched here in terms of four A and three B in the left, three A and five B in the center, and five A and four B in the right section at $t = 0$. With progressing reaction $t > 0$, the diffusion-controlled reactions between A and B, each A (B) *solely* reacting with another B (A), gradually enlarge linear size $\xi(t)$ of a concentration-unbalanced domain, as is demonstrated by the middle line of Fig. 4.1, where reactant A is predominant in the two outer and reactant B in the central section. Later, after all remaining $A + B \Rightarrow 0$ have reacted—recall that each A (B) strictly reacts with one B (A)—an extended local domain of reactant A survives from the homogenous, initial mixture: Winner takes all [19].

Therefore, we have to take account of the random fluctuations of the concentrations of reactants A and B—regardless of whether the system is a fractal. We know that for a random distribution of a large number N of reactant particles—as assumed here—there are local number fluctuations that increase or decrease the value of N by $N^{1/2}$, on average. Consequently, we express the number of reactant molecules $N_{A,B}$ within a *local* volume $V(\xi)$—scaling with distance ξ as $V(\xi) \sim \xi^{\vartheta}$ [see Eq. (3.5)]—as

$$N_{A,B} \sim [\rho_{A,B}(0)]\xi^{\vartheta} \pm \{[\rho_{A,B}(0)]\xi^{\vartheta}\}^{1/2} \sim [\rho_{A,B}(0)]\xi^{\vartheta} \pm [\rho_{A,B}(0)]^{1/2}\xi^{\vartheta/2} \quad (4.11a)$$

After (diffusion) time $t \approx \xi^2/K_0$ (see Section 4.1), each of the molecular species A and B has reacted with its counterpart (last line of Fig.4.1); the system is now a *homogenous* mixture of alternating A- and B-*rich* domains $V(\xi)$ of particle number $N_{A,B}(\xi) \sim \{[\rho_{A,B}(0)]\xi^{\vartheta}\}^{1/2}$ or particle density $\rho_{A,B}(\xi) \sim N_{A,B}(\xi)/V(\xi) \sim \{[\rho_{A,B}(0)]\xi^{\vartheta}\}^{1/2}/V(\xi) \sim \{[\rho_{A,B}(0)]\xi^{\vartheta}\}^{1/2}/\xi^{\vartheta} \sim [\rho_{A,B}(0)]^{1/2}\xi^{-\vartheta/2}$, each domain stretching over linear range $\xi \approx (t/K_0)^{1/2}$. Therefore, toward depletion of unreacted A and B, the *global* density of remaining reactant, say A, is about one-half the density it has within a single domain (cf. the second and third lines in Fig. 4.1):

$$\rho_A(\xi) \approx (\tfrac{1}{2})[\rho_A(0)]^{1/2}\xi^{-\vartheta/2} \sim [\rho_A(0)]^{1/2}t^{-\vartheta/4} \sim \rho_A(t) \qquad \rho_A(0) = \rho_B(0) \quad (4.11b)$$

($t = 0$):   ...|ABABAAB|BABBBABA|BABABAAAB|...

($t > 0$):   ...| A | B | A |...
                    <-----$\xi(t)$----->

          ...| A |...

Fig. 4.1 Schematic of reactant separation during the diffusion-controlled process $A + B \Rightarrow 0$ in a one-dimensional space. [Reprinted with permission from K. Kang and S. Redner, *Phys. Rev. A* **32**, 435 (1985), Fig. 2, American Physical Society.]

Equation (4.11b) tells us that reaction rate $-d\rho_{A,B}(t)/dt \sim (d/dt)t^{-\vartheta/4} \sim t^{-(\vartheta/4)-1}$ of reactant *density difference fluctuations* is slower than reaction rate $-d\rho_{A,B}(t)/dt \sim (d/dt)t^{-\vartheta/2} \sim t^{-(\vartheta/2)-1}$ of reactant *densities* [first term of Eq. (4.11a)]. Therefore, the decay rate of the density difference fluctuations due to diffusion becomes *rate determining*. Solving now for t in Eq. (4.11b) and inserting the result into rate $-d\rho_{A,B}(t)/dt \sim t^{-(\vartheta/4)-1}$ (see above) then yields

$$-d\rho_{A,B}(t)/dt \sim [\rho(t)]^{(\vartheta + 4)/\vartheta} \tag{4.11c}$$

As mentioned above, the phenomenon is quite general for diffusion-controlled $A + B \Rightarrow 0$ reactions and we must now extend these modifications to *fractal* substrates (under reentrant conditions, $\bar{d} < 2$). To accomplish this, may we simply substitute exponent ϑ by fracton dimension $\bar{d}$ as we did for reaction scheme $A + A \Rightarrow 0$ (see Section 4.2.1)? The answer is no. For instance, assume reactant B in a local domain of $B \approx A$ (see Fig. 4.1) approaches a local A cluster of volume Λ^d and surface Λ^{d-1}, where Λ is a radius of gyration and d the fractal dimension of the clusters. It is obvious that only the A molecules on the surface of A clusters encounter a diffusing B; those in the interior of the A clusters are screened off. The diffusive $A + B$ encounter frequency must be corrected by factor $\Lambda^{d-1}/\Lambda^d = \Lambda^{-1}$. The final result is [20]

$$\rho_A(t) \sim \rho_B(t) \sim \rho(t) \sim t^{-\gamma} \quad \gamma = (\bar{d}/2)(1 - \bar{d}/2d) \tag{4.12}$$

and not $\rho(t) \sim t^{-\bar{d}/4}$ [see Eq. (4.11b)]. Because $\bar{d} < d$ [Eq. (4.7)], it follows that $\gamma > \bar{d}/4$ and we conclude that reactant clustering indeed leads to a slower diffusion-controlled reaction rate of the density difference fluctuations in the fractal medium. However, because numerical differences between fracton and fractal dimensions are usually small (see Section 4.1), it requires very precise experimentation to ascertain variations of Eq. (4.12) from $\bar{d}/4$ behavior [Eq. (4.11b)].

Finally, we establish the rate expression under boundary condition $\rho_A(0) \neq \rho_B(0)$, say $\rho_A(0) < \rho_B(0)$. Recall that the mean-field expression for $\rho_A(0) \approx \rho_B(0)$, namely, $\rho_A(t) \sim K_0 t^{-1}$, was modified to $\rho_A(t) \sim t^{-\vartheta/4}$ [Eq. (4.11b)] in order to introduce the correct time dependence of the rate-determining density difference fluctuations. Assuming then that the applied correction, $t \Rightarrow t^{\vartheta/4}$, *also* holds under boundary condition $\rho_A(0) < \rho_B(0)$, we raise time t in the classical expression $\rho_A(t) \sim \exp\{-K_0[\rho_B(0) - \rho_A(0)]t\}$ to the power of $\vartheta/4$. This yields

$$\rho_A(t) \sim \exp[-K_0(c_B - c_A)t^{\vartheta/4}] \quad \rho_A(0) \neq \rho_B(0) \tag{4.13}$$

Equation (4.13) represents an _extended_ or _stretched exponential_, with $c_{A,B} = [\rho_{A,B}(0)]^{1/2}$ and $\vartheta = E$ (regular diffusion) or $\vartheta = \bar{d}$ (anomalous diffusion) [18].

Perhaps it is now helpful to summarize the contents of this section by showing computer-modeled results of $A + B \Rightarrow 0$ on the example of the relative densities of $\rho_A(t)/\rho_A(0)$ of remaining reactant A for various conditions of initial densities $\rho_A(0)$ of A and $\rho_B(0)$ of B. For the sake of simplicity, Fig. 4.2 displays appropriate results

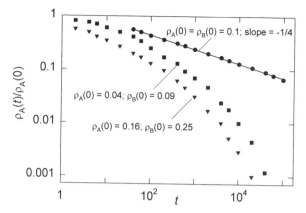

Fig. 4.2 Presentation (in decimal log–log coordinates) of the computer-modeled time dependence of the concentration ratio of nonreacted to initial densities for reagent A, undergoing diffusion-controlled reaction $A + B \Rightarrow 0$ with randomly distributed (fixed) B on a one-dimensional line. Three different sets of initial reactant densities $\rho_A(0)$, $\rho_B(0)$ are displayed. [Reprinted with permission from K. Kang and S. Redner, *Phys. Rev. A* **32**, 435 (1985), Fig. 1a, American Physical Society.]

of reaction $A + B \Rightarrow 0$ of isotropically diffusing A particles on a line lattice ($\vartheta = 1$) of 10^6 sites, with randomly distributed and position-fixed reactants B, and their subsequent removals upon $A + B$ encounters [18]. Boundary conditions are $\rho_A(0)/\rho_B(0) = 1$, $\rho_A(0)/\rho_B(0) = 0.44$, and $\rho_A(0)/\rho_B(0) = 0.64$, respectively, thus comparing the computer predictions with those of Eq. (4.11b) for system with $\rho_A(0) = \rho_B(0)$ or Eq. (4.13) for $\rho_A(0) \neq \rho_B(0)$.

The uppermost curve in Fig. 4.2 shows system $\rho_A(0) = \rho_B(0)$, with slope $s = -\frac{1}{4}$—as predicted by Eq. (4.11b) for $\vartheta = 1$: $\rho_A(t)/\rho_A(0) \sim t^{-\vartheta/4} \sim t^{-1/4}$. Note that computer data and theory (slope = $-\frac{1}{4}$) agree only at longer times, namely, when depletion of reactants A, B is approached and the density difference fluctuations begin to be rate determining.

The two lower plots in Fig. 4.2 display the results for systems $\rho_A(0) \neq \rho_B(0)$, with A as minority species. Here, the stretched exponential function [Eq. (4.13)] is relevant, as implied by the downward curvature of the computer data within the log $[\rho_A(t)/\rho_A(0)]$ − log (t) coordinate system.

Finally, Fig. 4.3 displays a replot of the data of Fig. 4.2 in terms of log(log $\{1/[\rho_A(t)/\rho_A(0)]\}$) versus log (t), which readily shows that computer results and theory agree at longer reaction times (slope = $\frac{1}{4}$).

4.2.3 Anomalous Diffusion in a Triblock Copolymer During Gelation

We had previously looked at static fractal aspects of sol–gel phenomena during size-exclusion chromatography (Section 3.1.1); we will now take up aspects of a gel as it influences dynamical phenomena taking place within it; in particular, we

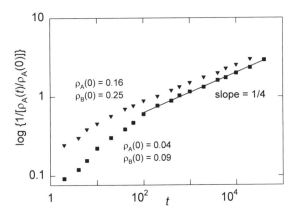

Fig. 4.3 Decimal log–log graph of log $\{1/[\rho_A(t)/\rho_A(0)]\}$ versus t from the data series for conditions $\rho_A(0) \neq \rho_B(0)$ of Fig. 4.2, emphasizing their stretched exponential time dependence at long times.

explore how irregular gel structures impede diffusion of a tagged particle. Such experimental data are of wider importance; flux and transport through irregular materials and on non-Euclidian surfaces is ever-present in chemical practice.

Specifically, we will discuss self-diffusion (by standard pulsed proton NMR techniques) of a commercial poly(ethylene oxide$_{78}$)–poly(propylene oxide$_{30}$)–poly(ethylene oxide$_{78}$) triblock copolymer, Pluronic F68™, of molecular weight $\approx$ 8000. The material, in a 35% wt aqueous solution, passes into its gel regime at temperatures exceeding about 37 °C [21]. Hence, it is an interesting object for studying the temperature dependence of its self-diffusion coefficients with the particular aim of ascertaining any anomalous behavior when the system passes from its mobile sol into the gel phase.

In brief, one observes the attenuation of the pulsed proton spin–echo signal Ψ with time due to the loss of translational correlation between the diffusing copolymer molecules. For regular diffusion, Ψ is a Gaussian distribution $\Psi(q^2\Delta) = \exp(-q^2\langle z^2 \rangle) \equiv \exp(-q^2\xi^2) = \exp(-q^2 K_0 \Delta)$ in terms of pulse width separation Δ (time) (effectively the "observation time"), scattering vector q (length^{-1}) [22], mean-squared translational displacement or variance $\langle z^2 \rangle \equiv \xi^2 = K_0 \Delta$ of the random displacement coordinate z, and effective self-diffusion constant K_0 (see Glossary 2, *random walk*). For anomalous polymer diffusion, we formally modify $\xi^2 = K_0 \Delta$ to $\xi^2 \sim K_0 \Delta^{2/D_W}$, $D_W > 2$, as suggested by Eq. (4.1).

Figure 4.4 shows plots of log (ξ^2) versus log (Δ) of the signal attenuation data for the system in its sol and gel states. For the sol (system below about 37 °C) we find slope $s = 2/D_W \approx 1$, indicating normal diffusion. However, once the system has moved into its gel zones (near 37 °C and above), the slope drops below unity, with decreasing values under increasing gel formation. Consequently, diffusion of the copolymer in its gel zones is anomalous; for instance, slope $= 0.56$ corresponds to $D_W = 3.6 \gg 2$. A numerical value of $D_W \approx 4$ has been measured at the gel point of

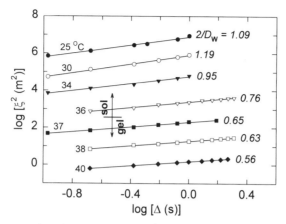

Fig. 4.4 Mean-squared random displacement ξ^2 of triblock copolymer F68 in its sol and gel zones as function of observation time interval Δ from its proton spin–echo attenuation data. System temperatures (in centrigrade) and slopes (exponent $2/D_W$) of log (ξ^2)/log (Δ) are indicated. [Reprinted with permission from H. Walderhaug and B. Nyström, *J. Phys. Chem. B* **101**, 1524 (1997), Fig. 3, Copyright 1997 American Chemical Society.]

other polymeric systems and fits predictions from computer models and well-established theories on diffusion in percolating systems [3].

To specify a fracton dimension, we take fractal dimension $d \approx 2.5$ of percolation or DLA clusters in $E = 3$ (see Section 3.1.2) and therefore predict spectral dimension $\bar{d} \approx 1.4$ for gel-like copolymer self-diffusion with the help of Eq. (4.6) and numerical value $D_W = 3.6$. This estimate is close to the fracton dimension of the incipient percolation cluster in $E = 3$ ($\bar{d} = 1.42$) [8, 16].

Wave vector probing of the spin–echo measurements at different pulse widths Δ indicates that exponent $2/D_W$ for anomalous transport of the copolymer in its gel state is weakly q dependent: The data, plotted in Fig. 4.5 for two temperatures above the sol–gel point, show that $2/D_W$ decreases with increasing q, the decrease being smaller at higher temperatures. Recalling that q^{-1} is the spatial resolution width of the experiment (see Section 3.1.2.2), smaller exploration volumes are monitored at higher q. On the other hand, a smaller exploration volume translates to greater displacement restrictions. Hence, the larger the scattering wave vector q, the smaller the extent of diffusing space we observe, and the greater the diffuser's delay by the system's connectivity [10]. Because larger q relate to increased compactness of the random walk, exponent $2/D_W$ should decrease with increasing q, as indeed shown in Fig. 4.5 [21].

4.2.4 Diffusion-Controlled Reactant Density Fluctuation Kinetics: Hydrogen Exchange in Lysozyme–Water

Analysis of surface properties of the globular proteins lysozyme, myoglobin, trypsin, and subtilisin indicates fractal surface dimensions $d = 2.1$–2.2. These values do not

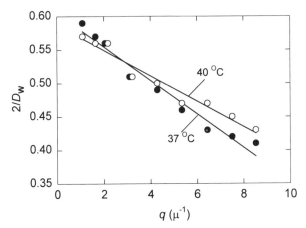

Fig. 4.5 Scattering wave vector dependence of exponent $2/D_W$ of the (mean-squared) displacement–time scaling relation of self-diffusion of the F68 triblock copolymer of Fig. 4.4 at incipient gel formation (near 37 °C) and within its gel zones (40 °C). [Reprinted with permission from H. Walderhaug and B. Nyström, *J. Phys. Chem. B* **101**, 1524 (1997), Fig. 4, Copyright 1997 American Chemical Society.]

depend on the origin of the proteins, their molecular mass, molecular diameter, or—of greatest interest—their biological function [23]. Consequently, the dynamic aspects of such a biological macromolecule are indeed interesting if not vexing, as shown by the diffusion-limited proton exchange between solvent water and tritium-exchanged lysozyme [24]. The system, prepared under replacement of all its exchangeable protons on the protein by the tritium label, is kept at sufficiently low temperatures to block its exchange with the water protons, eventually abruptly heated to temperatures up to 56.5 °C to induce exchange, and followed by standard analysis of the retained amount of proton label at a range of reaction times. Previous classical interpretations of the global kinetics [25] had indicated an exceptionally broad distribution of individual exchange rates involving most of all exchangeable hydrogen, making it impossible to obtain a singular reaction rate constant.

Because the labeled hydrogen density $\rho_A(t)$ on the protein is the minority species, $\rho_A(0) \ll \rho_B(0)$, we use the stretched exponential function $\rho_A(t)/\rho_A(0) \sim \exp[-(t/\tau)^{\vartheta/4}]$ [Eq. (4.13)] to evaluate the experimental results of the time-dependent exchange of lysozyme-bound tritium by water protons [24–26]. Figure 4.6 displays its log plot (decimal logarithm) of nonexchanged labeled protons versus abscissa coordinates $t^{\vartheta/4}$ for aqueous lysozyme in the presence of inhibitor (*N*-acetyl-D-glucosamine) at the indicated temperatures in degree Celsius. Notice that the least-squares slopes are linear under the same value of $\vartheta/4 = 0.21$ [24]. Finally, we relate the experimental slopes s of Fig. 4.6 to inverse relaxation times $1/\tau$ by relation $s/\log(e) = -(1/\tau)^{\vartheta/4}$, with $e = 2.718$; the usual activation plot of $1/\tau = \exp(-\Delta E/kT)$, in terms of $\ln(1/\tau)$ versus $1/T$ (K), is displayed in Fig. 4.7.

How can we interpret dimension $\vartheta \approx 0.84$? Briefly, it is proposed [24] that

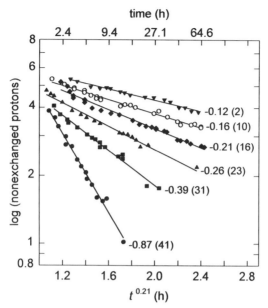

Fig. 4.6 Time dependence of labeled proton exchange on lysozyme with the protons of its 20-mg/l aqueous solution at pH 6.8 in the presence of 0.2 M N-acetyl-D-glucosamine inhibitor. The abscissa is in units of $t^{0.21}$, the ordinate in decimal logarithm of labeled proton retention on the protein. The straight lines are linear least-squares plots; standard deviations are, at worst, 7% of the value of a slope. Temperatures (°C) are indicated within parentheses. [Reprinted with permission from T. G. Dewey, *Fractals* **3,** 251 (1995), Fig. 2, World Scientific Publishing Co.]

$\vartheta \approx 0.84$ represents the dimension of the reaction volume or boundary volume of the fractal protein surface, namely, $d = 3(\text{solution}) - 2.17(\text{lysozyme surface}) = 0.83$. The boundary volume is, so to speak, the complementary space of the fractal protein surface, $0.83 + 2.17 = 3$.

The low value of dimension ϑ of the reaction volume defined above should not be a surprise: We have seen in Section 4.2.2 that fluctuations of reactant concentrations play a dominant role at longer reaction times and $\vartheta = 0.84$ is the concordant result [Eq. (4.13)]. What this all means in terms of a definite mechanistic-quantitative picture of the reaction steps between tritium label on the protein and protons in the surrounding aqueous medium will probably require additional experimental efforts.

4.2.5 Sintering of Open Sites on Dispersed Platinum

Figure 4.8 displays a log–log replot of originally linearly presented data of the time dependence of deactivation of a dispersed Pt catalyst used in an industrial carbohydrate oxidation process [27]. The implication here i ⸱hat temperature-induced

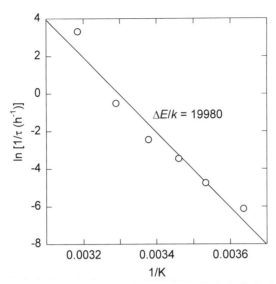

Fig. 4.7 Activation plot of rate constants $1/\tau$ computed from the least-squares slopes of Fig. 4.6. [Reproduced with permission from T. G. Dewey, *Fractals* **3**, 251 (1995), Fig. 3, World Scientific Publishing Co.]

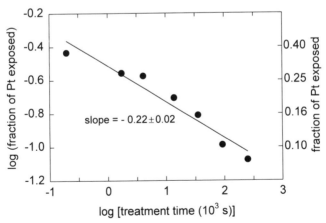

Fig. 4.8 Decrease of exposed fraction of Pt atoms of a fresh graphite-supported Pt catalyst with duration of a sintering treatment in 0.1 M NaOH under H_2 at 363 K. The fractional area, obtained by CO chemisorption, is plotted log–log (decimal) versus treatment duration. [Reproduced with permission from J. H. Vleeming, B. F. M. Kuster, G. B. Marin, F. Oudet, and P. Courtine, *J. Catal.* **166**, 148 (1997), Fig. 5, Academic Press, Inc.]

sintering by diffusion-limited aggregation is the cause of the deterioration of catalyst open-site density. Note that slope $s = -0.22$ suggests an A + B ⇒ 0 process running under dimensional value of $\vartheta \approx 0.9$ [Eq. (4.11b)], perhaps with A and B representing local, active Pt clusters of different average number density diffusing toward each other on essentially one-dimensional spaces to form less open Pt-surface regions. Figure 4.9 [27] displays the image of a particularly large Pt aggregate formed during the sintering process. Its shape, indicating loops and dead-end branches, is quite reminiscent of the structural patterns of the random-walk aggregation clusters we discussed previously.

4.2.6 A Summary

The value of theoretical predictions of anomalous diffusion and fractal reaction kinetics to guide laboratory research is unquestionably beneficial. Among other aspects, we have learned that fluctuations of the reactants are important—something most chemists would not, but should, worry about if a batch reaction under diffusion-controlled conditions is carried to higher percentages of conversion. Furthermore, it is relatively easy to control input parameters and boundary conditions in computer models (well, not always: Edward Lorenz had something to say about this); therefore, all kind of situations can be played through, which—in my opinion—explains an obviously large preponderance of computer-model over laboratory studies. For instance, a model of the diffusion-limited process A + B ⇒ 0 has been explored under *correlated* initial A, B distributions; reactant pairs A, B are separated by some distance [28]. The model shows that for recurrent (compact) walk, exponent $\vartheta/4$ in $\rho(t) \sim t^{-\vartheta/4}$ [Eq. (4.11b)] is to be modified to $(\vartheta + 2)/4$, not an insignificant change.

Calculations of positional correlation data by exact enumeration are similarly

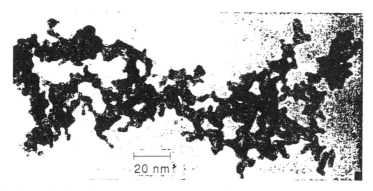

Fig. 4.9 Transmission electron microscopic image of a large Pt aggregate formed during reductive treatment of a fresh graphite-supported Pt catalyst for 70 h (see Fig. 4.8). [Reprinted with permission, J. H. Vleeming, B. F. M. Kuster, G. B. Marin, F. Oudet, and P. Courtine, *J. Catal.* **166,** 148 (1997), Fig. 6, Academic Press, Inc.]

useful for checking the range of validity of scaling relations. For instance, estimating spectral dimension $\bar{d}$ on a DLA cluster from the velocity autocorrelation of a random diffuser yields, for $E = 3$ embedding space, spectral dimension $\bar{d} = 1.31 \pm 0.02$ and parameter $D_{\mathrm{w}} = 3.21 \pm 0.04$ [29]. Recalling general scaling relation Eq. (4.6), $\bar{d} = 2d/D_{\mathrm{w}}$, and fractal DLA dimension $d = 2.48 \pm 0.02$ (see Section 3.1.2), we now note a serious inconsistency between the two outcomes: Equation (4.6) predicts $\bar{d} = 1.54 \pm 0.01$, a value about 15% higher than $\bar{d} = 1.31 \pm 0.02$ from the enumeration method.

It seems that two simultaneous conditions cause this breakdown of scaling equation Eq. (4.6) for the DLA cluster: The walker is trapped in local regions of the cluster and the fractal dimension of these local regions is different from that of the cluster [29]. Clearly, such experimental outcomes would be difficult to understand without knowledge or, at least, awareness of the underlying fracton theories.

It is now useful to provide an operational definition of the fracton for practitioners: At the end of Section 3.2.3, a fractal was characterized as an object resulting from a random iterative, optimally space-filling process under some system constraints. Here, we may find the following useful:

> A fracton is a number characterizing the progress of random transport of some quantity over a fractal substrate under the constraints of the latter's connectivity, its branching/looping patterns.

4.3 THE REACTION DIMENSION

4.3.1 Particle and Active Site Effects in Catalyzed Reaction Rates

Not all chemical processes obey diffusion-controlled mechanisms, not all chemical reactions are carried to depletion of reactant nor do they occur without back-reaction steps and in the absence of catalysis; technological and efficiency considerations toward a desired economical outcome or choice of product dictate the boundary conditions, reaction schemes, and makeup of the system. In other words, the reaction proceeds under mean-field theory conditions—briefly mentioned at the begin of Section 4.2.2: (a) Fluctuations of reactant density differences are neglected as immaterial and (b) reaction rates or conversion yields are counted after initial transients have dissipated.

These two aspects complicate a successful application of fracton theory in its ultimate goal, namely, to yield information on the structural pattern of the fractal medium over which the particular transport process occurs, because the *local* connectivity appears "washed out" to the probing mechanism [30]. Yet, it has turned out that a general property–distance power law approach of the type of Eq. (2.17) accounts for many aspects of chemical reaction rates that involve some solid phase of known particle size R, such as catalytic processes and dissolution phenomena. We recall the usefulness (as well as restrictions) of scaling chemisorption capacities m_{C} with chemisorption sites in terms of chemisorption dimension D_{C} as exponent, $m_{\mathrm{C}} \sim R^{D_{\mathrm{C}}}$ (see Section 2.4.1). For purposes of reaction dynamics, we retain the same power law formalism but consider as dependent property the measured reactivity, turnover rate, or other convenient signifiers of chemical reactions, a_R, and replace chemisorption dimension D_C by reaction dimension D_R [31],

$$a_R \sim R^{D_R} \tag{4.14}$$

We shall understand that D_R does not necessarily imply a fractal or a fracton dimension in their strict sense; we are reminded here of the parallel situation discussed in Section 2.4.4 for chemisorption dimension D_C: Essentially, D_R quantifies how the chemical reactivity of some reaction increases (or decreases) with a linear increase of the size of the surface (or solid) on which the reaction takes place, or by which it is catalytically accelerated, or otherwise participates within the overall rate-determining scheme. Note, then, that Eq. (4.14) numerically evaluates how a chemical rate a_R depends on *active* site properties, with obvious implications to structure sensitivity, product selectivity, site activity under catalyst loading, sintering, poisoning, and pretreatment or in situ induced modifications of site activity [31].

Frequently, chemical activity a_R is reported in units reduced to mass (volume) or to surface area, in which case Eq. (4.14) is modified, respectively, to

$$a_R \sim R^{D_R}/R^3 \sim R^{D_R-3} \tag{4.15a}$$

$$a_R \sim R^{D_R}/R^2 \sim R^{D_R-2} \tag{4.15b}$$

4.3.1.1 Ethylene Epoxidation over Dispersed Silver

Reaction data of the Ag catalyzed epoxidation of ethylene [32] are now used to explore evaluation of dimension D_R; Figure 4.10 shows the appropriate plot of log (a_R) versus log (R) from the tabulations of reaction rate a_R (in molecule $s^{-1}m^{-2}$) and average Ag particle size R (Å) of the α-Al$_2$O$_3$-supported catalyst. Hence, slope s of the plot equals $D_R - 2$ [Eq. (4.15b)]. We note: (a) For Ag particles up to about 500 Å, the reactivity of the active sites increases strongly with particle size R: $D_R \approx 3.4 + 2 = 5.4$. (b) At particle sizes exceeding 500 Å, further growth in a_R is much lessened: $D_R \approx 2.6$.

What does this tell us? What could be the significance of an exponent of $D_R \approx 5$ obtained from a power law scaling relation? Certainly, such value cannot represent a *geometrical* dimensional quantity pertaining to the supported Ag crystallites; indeed, size determination had established that the metallic crystals exhibit Euclidian properties [33], with surface dimension $d \approx 2$. Consequently, $D_R \approx 5$ refers to a property in the overall catalyst reactivity that observably grows much faster with increasing crystal size than predicted by purely geometric considerations.

The observed reactivity dependence on crystal size is associated with the presence of catalytically active intergrain boundaries of the Ag crystals [32], implying that formation of such structure faults is favored by Ag crystal growth (larger crystals possess a higher probability to develop structure faults). Furthermore, numerical value of reaction dimension $D_R = 2.6$ predicts that the relative increase in the chemical activity, $(1/R)da_R/dR \sim R^{2.6-2} \sim R^{0.6}$, has dropped by a factor of $R^{3.4}/R^{0.6} \approx R^3$ compared to its growth at $R < 500$ Å. Consequently, employ of Ag

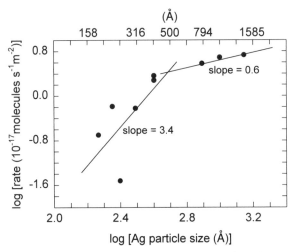

Fig. 4.10 Log–log (decimal) scaling plot of the reaction rate of ethylene epoxidation on an Ag/α-Al$_2$O$_3$ catalyst versus average Ag particle size from transmission electron microscopy. The straight lines represent the linear least-square computations through the first six and last three data points, respectively. [Data from S. V. Tsybulya, G. N. Kryukova, S. N. Goncharova, A. N. Shmakov, and B. S. Bal'zhinimaev, *J. Catal.* **154**, (1995), Tables 1 and 2.]

crystals exceeding 500 Å would offer no advantage to the overall economy of the process.

Keep in mind that all this has nothing to do with a fractal *surface* dimension.

4.3.1.2 *Oxidative Dehydrogenation of Propane*

Next, we discuss aspects of active site distributions of vanadium on vanadium–niobium catalysts in the oxidative dehydrogenation of *n*-propane, with emphasis on catalytic surface site activities as probed by low-energy ion scattering (LEIS) [34, 35]. Because we cannot credibly estimate catalyst particle size R from the catalyst surface concentration C_S of the LEIS results, we assign scale of Eq. (4.14) to C_S. Hence, we write $a_R \sim C_S^{D_R}$.

Figure 4.11 shows the logarithm of the intrinsic activity a_R of propane consumption at 425 °C versus the logarithm of the analytical surface vanadium concentration C_S of the V–Nb catalyst, replotted from the linear presentation of the data. Least-squares slope $D_R = 0.67 \pm 0.17 < 1$ signifies that intrinsic site activity a_R for oxidative propane consumption increases slower than the analytical surface vanadium concentration. The proportion of *active* vanadium surface sites rises less rapidly than the number of *total* vanadium surface sites.

It is suggested [34] that active vanadium surface sites on the catalyst are distributed in the form of vanadium atoms and their two- and three-member aggregates, with greater activity ascribed to the clusters. However, we can argue that (a) if only

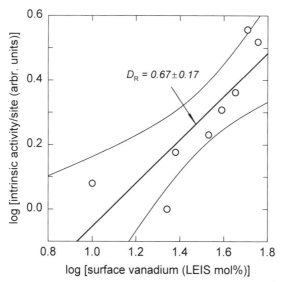

Fig. 4.11 Log–log (decimal) plot of the intrinsic activity of propane consumption per vanadium surface site during oxidative dehydrogenation of propane over a V–Nb catalyst as function of its surface V concentration by LEIS. Least-squares slope and 95% confidence limits are shown. [Reproduced with permission from R. H. H. Smits, K. Seshan, J. R. H. Ross, L. C. A. van den Oetelaar, J. H. J. M. Helwegen, M. R. Anantharaman, and H. H. Brongerma, *J. Catal.* **157**, 584 (1995), Fig. 7, Academic Press, Inc.]

three-member vanadium aggregates are active, D_R should equal ⅓, (b) if solely two-member clusters are active, numerical value $D_R = ½$ is predicted, and (c) if only single-atom sites are active, $D_R = 1$ should be found. Therefore, the observed $D_R = 0.67$ implies that single-atom sites are participating in the reaction to a *considerable* extent.

Note again that the obtained result has nothing to do with any fractality of the proper catalyst surface; the evaluations deal solely with the distribution of reactive sites during a specific process under given reaction conditions of temperature and types of reagent, and for selected dynamic and static quantities. The geometrical substrate and its unspecified dimensionality are unimportant here.

4.3.1.3 Nitrobenzene Hydrogenation over Platinum Dispersed on Carbon Cloth

Figure 4.12 shows the log–log replot of linear data of specific activity of nitrobenzene reduction to aniline and phenylhydroxylamine over Pt dispersed on carbon cloth [36] at four different degrees of metal dispersion (from H_2 chemisorption). Because a_R is reported in units of mmol min^{-1}g^{-1}, slope $s = 0.955$ is, in principle, related to D_R by $s = D_R - 3$ [Eq. (4.15a)]. But as the Pt particle size data are given here as Pt dispersion rather than Pt particle diameter R, we must substitute slope s by its negative value. Particle dispersion is in-

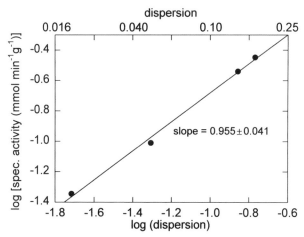

Fig. 4.12 Plot of the decimal log (activity) of nitrobenzene consumption versus log (Pt dispersion) for the hydrogenation reaction of nitrobenzene to aniline and phenylhydroxylamine at 323 K. [Reproduced with kind permission from M. C. Marcías Pérez, C. Salinas, M. de Lecea, and A. Linares Solano, *Appl. Catal. A* **151**, 461 (1997), Fig. 5, Elsevier Science-NL, Sara Burgerhartstraat 25, 1055 KV Amsterdam, the Netherlands.]

versely proportional to particle radius R or $R \sim 1/\text{dispersion}$ [37]. With $s = -0.955$, we thus obtain $D_R = 2.05$.

The result shows quantitatively that the Pt–carbon cloth catalyzed reaction of nitrobenzene reduction to aniline and phenylhydroxylamine is *structure insensitive* [36]: A change in the number of *analytical* Pt sites engenders the identical change in the number of *active* Pt sites [31].

4.3.1.4 Effect of Platinum Poisoning by Electrolytic Sulfur Deposition on Maleic Acid Hydrogenation

In this section, we deal very briefly with effects of sulfur poisoning, by electrodeposition, of a Pt catalyst prior to its use in the hydrogenation of maleic acid [38]. We had previously discussed some fractal aspects of diffusion-controlled electrolytic metal deposition (Section 3.2.1). Hence, it would be interesting and useful to explore specific surface effects leading to blockage of active sites caused by such methods.

Figure 4.13 displays two log–log scaling plots of the relative intrinsic activity a_R/a_0 of maleic acid hydrogenation at concentrations 2.5×10^{-3} (▼) and 2.5×10^{-2} M (●) in water, respectively, versus the degree of unblocked (nonpoisoned) Pt sites, $1 - \theta_S$, under prior electrodeposited sulfur coverage θ_S. In the lower concentration system $(2.5 \times 10^{-3}$ M), a_R/a_0 scales with $1 - \theta_S$ as $a_R/a_0 \sim (1 - \theta_S)^n$, where $n = D_R = 6.6 \pm 1.2$ from Eq. (4.14). For the 2.5×10^{-2} M system, the reaction dimension amounts to $D_R = 1.37 \pm 0.07$.

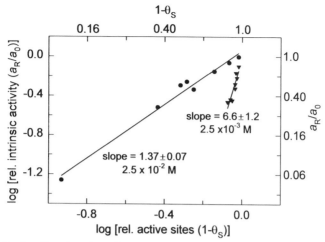

Fig. 4.13 Decimal log–log plots of the intrinsic activity versus fraction $1 - \theta_S$ of unblocked sites for maleic acid hydrogenation at two concentrations over a Pt catalyst poisoned with electrodeposited sulfur to relative coverage θ_S. Least-squares linear slopes and their standard deviation are indicated. [Reproduced with kind permission from E. Lamy-Pitara and J. Barbier, *Appl. Catal. A* **149**, 49 (1997), Figs. 19 and 20, Elsevier Science-NL, Sara Burgerhartstraat 25, 1055 KV Amsterdam, the Netherlands.]

What do these results signify? Consider that $a_R/a_0 \approx (1 - \theta_S)^7$, with reaction dimension $D_R \approx 7$ for the most dilute maleic solutions (▼), yields the probability $P_7 = P_1^7$ of finding an active site consisting of 7 unblocked sites, each with probability $P_1 = 1 - \theta_S$. Similarly, $D_R \approx 1.4$ implies the corresponding probability for the 10-fold more concentrated system of maleic acid (●). Consequently, the *active* site for maleic acid hydrogenation consists of about 7 unblocked Pt sites for the 2.5×10^{-3} M, or 1–2 unblocked (nonpoisoned) Pt sites for the 2.5×10^{-2} M system [38].

This finding is quite remarkable. For instance, from Fig. 4.13 we take that at common value $a_R/a_0 = 0.5$, the relative active Pt site requirements for the 2.5×10^{-3} M system are 0.91 compared to 0.55 for the 2.5×10^{-2} M system, the difference becoming more pronounced at lower a_R/a_0.

4.3.1.5 Influence of Pore Volume on Cracking Behavior of Zeolites
In Section 2.3.2, we discussed fractal aspects of gas adsorption on extraframe zeolite sites generated during standard zeolite treatment procedures. Here we explore scaling of the cracking activity of various zeolites with their void spaces.

Figure 4.14 shows a log–log plot of the product ratio of *iso*-C_4 to total C_4 versus zeolite void space during a cracking process of *n*-heptane [39] for various zeolites named in the body of the figure; the original linear data are shown as inset. Specifically, zeolite void space is given in terms of the diameter of channels

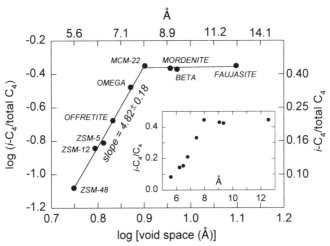

Fig. 4.14 Decimal log–log plot of product ratio *iso*-C_4/total C_4 during cracking of *n*-heptane over several zeolites as function of their void space. The inset shows the linearly plotted data. [Reprinted with permission from A. Corma, M. Davis, V. Fornés, V. González-Alfaro, R. Lobo, and A. V. Orchillés, *J. Catal.* **167**, 438 (1997), Fig. 6, Academic Press, Inc.]

of unidirectional zeolite, space generated by the crossing points of intersecting channels in multidirectional zeolite, and other zeolite cavities.

The scaling plot accentuates attainment of constant dependence of cracking activity on zeolite void space at diameters $R_m \geq 8$ Å, which seems to be a saturation point of activity increase, perhaps a minimum diameter needed to bring about sufficient reactant site efficiency. Note here that the cracking reactivity scales with void space under reaction dimension $D_R = 4.8$, a value that exceeds any purely geometrical requirements ($D_R \leq 3$) by, at least, $R^{1.8}$. Perhaps the sites that are active for the cracking process are composed of several zeolite sites; according to numerical value $D_R \approx 5$ this would amount to about 5 zeolite sites per active site (see Section 4.3.1.4). These results furnish another example where the reaction activity scaling approach points at active site distributions, with implications to structures of activated complexes of the reactants.

4.3.2 Particle Size Effects in Monomer–Polymer Interactions: Vinyl Chloride and Polyvinyl Chloride

Polyvinyl chloride (PVC) is of widest distribution in consumer-oriented applications such as food packaging and pipes for liquid and gaseous transport. Its geometric surface properties are therefore of more than parochial interest, particularly if they are probed by its monomer, vinyl chloride [40].

In brief, the PVC surface measurements utilized the inverse surface tiling method (see Section 2.2) of PVC beads, of average size 90, 135, and 213 μm, by the vinyl chloride monomer molecule as its surface probe. Each PVC bead-size fraction was used as packing material in an inverse gas chromatography setup, the column being charged with the gaseous monomer.

The data, obtained in the temperature range 15–40 °C, are conveniently plotted as logarithm of the specific monomer retention volume per unit weight of packing (ml/g) versus the logarithm of PVC bead size. Consequently, the appropriate activity plot is given by $a_R \sim R^{D_R-3}$ [Eq. (4.15a)] yielding values of D_R between 2.87 and 2.95 at 15 °C and $D_R = 2.65$ at 40 °C.

Designating D_R as "fractal dimensionality of vinyl chloride monomer adsorption" [40] is most likely incorrect. First, interaction between these polar molecules is not negligible. Probably D_R more aptly characterizes a weak chemisorption state [41]—considering the column temperatures 15–40 °C—and is not a true measure of a surface dimension by a physisorption process. Second, the size range of the PVC beads covered a factor of merely 2.4, a range too narrow for drawing credible conclusions from data plotted in log–log coordinates.

4.3.3   A Summary

The five subsections under Section 4.3.1, dedicated to effects of particle size in chemical reactivity, should be considered to be principally of an advisory nature; the selected examples are meant to reemphasize the original idea and the significance of reaction dimension D_R [31]. I am pretty certain that many readers will be persuaded that scaling plots of chemical reactivity, as discussed here, will offer considerable and quantitative help in answering questions about the nature and distribution of active reaction sites—information that is readily obtained by extending standard analysis of physical catalyst particle size to a wider range than usually reported or required. Even rougher catalyst size distribution data, such as dispersion ranges obtained from chemisorption measurements, can be used for such purposes.

The scaling methods of chemical activity versus particle size are of general and predictive nature and provide additional insight, clearer theoretical guidance, and accordingly firmer conclusions. Whether the underlying size distribution is indeed fractal is not, a priori, guaranteed by the observed power law dependence of the appropriate quantities. However, statistical self-similarity of a reaction coordinate under finer and finer detail of the scale of its active site distribution very likely prevails in many chemical reactions involving amorphous-irregular solid surfaces [31].

NOTES AND REFERENCES

[1] This pictorial demonstration should be taken as such; it is very hard to combine introductory simplicity with theoretical precision; aspects of the latter will be discussed at length in this chapter.

[2] See [4] and [5] in Section 1.2.

[3] S. Havlin and D. Ben-Avraham, Diffusion in disordered media, *Adv. Phys.* **36**, 695 (1987).

[4] P. Evesque, *Energy Migration: Theory,* in *The Fractal Approach to Heterogeneous Chemistry,* D. Avnir, Ed. (Wiley, Chichester, UK, 1989), p. 93.

[5] See Eq. (16) of [52] in Section 2.4.3.

[6] From the definition of diffusion coefficient $K = \partial/\partial t\{\langle R^2 \rangle\}$ one gets [see Eq. (4.2)] $\partial/\partial t\{\langle R^2 \rangle\} = K_0(1 - \eta/2)t^{-\eta/2} = K(t)$.

[7] Quantity $p(t, \mathbf{r}_0 = 0) = (4\pi K_0 t)^{-E/2}$ is just the normalization constant of the normal or Gaussian probability distribution (see p. 28 in [12] of Section 1.2 and Glossary 2, *random walk*).

[8] See [33] in Section 3.2 and S. Alexander, Fractons, *Physica* **104A,** 397 (1986).

[9] R. Burioni and D. Cassi, Fractals without anomalous diffusion, *Phys. Rev. E* **49,** R1785 (1994); Spectral dimension of fractal trees, *Phys. Rev. E* **51,** 2865 (1995). The reports show a model of special tree-like fractals with fracton dimensions $\bar{d} > 2$ and noninteger, $\bar{d} = d$, hence $D_W = 2$ (regular diffusion).

[10] According to Merriam Webster's Collegiate Dictionary (10th ed., 1996), connectivity is "the quality or state of being connective or connected."

[11] This has been proven for $\bar{d}$ of infinite, *branched* irregular structures under the same nearest-neighbor adjacency relations. See D. Cassi and S. Regina, Spectral dimension of branched structures: Universality in geometrical disorder, *Phys. Rev. Lett.* **70,** 1647 (1993).

[12] Every herring is a fish, but every. . . . . .

[13] Dimension $E < 2$ of the embedding Euclidian space requires that $\bar{d} < 2$; see Eq. (4.7).

[14] R. Kopelman, P. W. Klymko, J. S. Newhouse, and L. W. Anacker, Reaction kinetics on fractals: Random-walker simulations and exciton experiments, *Phys. Rev. B* **29,** 3747 (1984); L. W. Anacker and R. Kopelman, Fractal chemical kinetics: Simulations and experiments, *J. Chem. Phys.* **81,** 6402 (1984).

[15] This is easily shown by differentiation of Eq. (4.9).

[16] The incipient percolation cluster (just at and slightly above the percolation transition) is a fractal. Further beyond the transition point, continual reaction of remaining finite clusters with the percolation cluster forms a Euclidian object (see also Section 3.1.1).

[17] D. Toussant and F. Wilczek, Particle-antiparticle annihilation in diffusive motion, *J. Chem. Phys.* **78,** 3642 (1983).

[18] K. Kang and S. Redner, Scaling approach for the kinetics of recombination processes, *Phys. Rev. Lett.* **52,** 955 (1984); Fluctuation-dominated kinetics in diffusion-controlled reactions, *Phys. Rev. A* **32,** 435 (1985).

[19] R. Kopelman, Fractal reaction kinetics, *Science* **241,** 1620 (1988), gives a colored graph of A, B reactant separation under diffusion-controlled reaction A + B over a percolating cluster grown on a two-dimensional lattice.

[20] W.-S. Sheu, K. Lindenberg, and R. Kopelman, Scaling properties of diffusion-limited reactions, *Phys. Rev. A* **42,** 2279 (1990).

[21] H. Walderhaug and B. Nyström, Anomalous diffusion in an aqueous system of a poly(ethylene oxide)–poly(propylene oxide)–poly(ethylene oxide) triblock copolymer during gelation studied by pulsed field gradient NMR, *J. Phys. Chem. B* **101,** 1524 (1997).

[22] Scattering vector q, of dimension length^{-1}, is the product of gyromagnetic ratio of ^{1}H, width, and amplitude of the gradient pulses.

[23] C.-D. Zachmann, S. M. Kast, A. Sariban, and J. Brickmann, Self-similarity of solvent-

accessible surfaces of biological and synthetic macromolecules, *J. Comput. Chem.* **14,** 1290 (1993); B. A. Fedorov, B. B. Fedorov, and P. W. Schmidt, An analysis of the fractal properties of the surfaces of globular proteins, *J. Chem. Phys.* **99,** 4076 (1993). The authors write ". . . fractal dimensions *d* of the protein surfaces . . . between 2.10 and 2.17 and thus were not much greater than the value *d* = 2 characteristic of a smooth surface." To this, refer to comments near the end of Section 2.3.4.

[24] T. G. Dewey, Chemically-controlled reaction kinetics on fractals: Application to hydrogen exchange in lysozyme, *Fractals* **3,** 251 (1995). There seems to be a mixup in the labeling of Figs. 1 and 2—it is Fig. 2 that presents the kinetic data in the presence of inhibitor. Also, the coordinate ranges of the presentations do not agree with the original data in [25] (see below). I have used what appears to be the correct versions.

[25] R. R. Wickett, G. D. Ide, and A. Rosenberg, A hydrogen-exchange study of lysozyme conformation changes induced by inhibitor binding, *Biochem.* **13,** 3273 (1974).

[26] Should we not modify the exponent in Eq. (4.13) in terms of Eq. (4.12)? Probably not (and nothing contrary has been found in the literature): Consider that species A was taken as minority reactant, $\rho_A(t) \ll \rho_B(t)$. Hence, a diffusing reactant A will readily encounter reactants B, regardless of the latter's clustering.

[27] J. H. Vleeming, B. F. M. Kuster, G. B. Marin, F. Oudet, and P. Courtine, Graphite-supported platinum catalysis: Effects of gas and aqueous phase treatments, *J. Catal.* **166,** 148 (1997). The catalyst was used in the oxidation of methyl α-D-glucopyranoside to Na-1-*o*-methyl α-D-glucuronate with molecular oxygen in a solution of 0.1 M aqueous NaOH.

[28] K. Lindenberg, B. J. West, and R. Kopelman, Diffusion-limited A + B ⟹ 0 reaction: Correlated initial condition, *Phys. Rev. A* **42,** 890 (1990).

[29] D. J. Jacobs, S. Mukherjee, and H. Nakanishi, Diffusion on a DLA cluster in two and three dimensions, *J. Phys. A* **27,** 4341 (1994).

[30] A. Blumen and G. H. Köhler, Reactions in and on fractal media, *Proc. Royal Soc. London Ser. A* **423,** 189 (1989); see the discussion on p. 199.

[31] D. Farin and D. Avnir, The reaction dimension in catalysis on dispersed metals, *J. Am. Chem. Soc.* **110,** 2039 (1988); see p. 286 of [45] cited in Section 2.4.1.

[32] S. V. Tsybulya, G. N. Kryukova, S. N. Goncharova, A. N. Shmakov, and B. S. Bal'zhinimaev, Study of the real structure of silver supported catalysts of different dispersity, *J. Catal.* **154,** 194 (1995).

[33] See [43] in Section 2.4.1.

[34] R. H. H. Smits, K. Seshan, J. R. H. Ross, L. C. A. van den Oetelaar, J. H. J. M. Helwegen, M. R. Anantharaman, and H. H. Brongerma, A low-energy ion scattering (LEIS) study of the influence of the vanadium concentration on the activity of vanadium–niobium oxide catalysts for the oxidative dehydrogenation of propane, *J. Catal.* **157,** 584 (1995).

[35] Low-energy ion scattering is sensitive to the first atomic layer of the material.

[36] M. C. Macías Pérez, C. Salinas, M. de Lecea, and A. Linares Solano, Platinum supported on activated carbon cloths as catalyst for nitrobenzene hydrogenation, *Appl. Catal. A* **151,** 461 (1997).

[37] The dispersion of a material on some substrate is inversely proportional to its average particle size. Imagine cutting a crystal successively into smaller pieces: The larger the dispersion of the ensemble, the smaller the average radius of a crystal. As concerns the scaling relation, notice that $a_R \sim R^{D_R-3}$ implies $a_R \sim dispersion^{-(D_R-3)}$.

[38] E. Lamy-Pitara and J. Barbier, Platinum modified by electrochemical deposition of adatoms, *Appl. Catal. A* **149,** 49 (1997).

[39] A. Corma, M. Davis, V. Fornés, V. González-Alfaro, R. Lobo, and A. V. Orchillés, Cracking behavior of zeolites with connected 12- and 10-member ring channels: The influence of pore structure on product distribution, *J. Catal.* **167,** 438 (1997).

[40] P. G. Demertzis and P. J. Pomonis, Fractal dimensionality of vinyl chloride monomer adsorption on polyvinyl chloride particles, *J. Coll. Interface Sci.* **186,** 410 (1997).

[41] See p. 282 of [45] cited in Section 2.4.1.

5

SPECTROSCOPY OF FRACTAL SYSTEMS

5.1 INTRODUCTION: GENERAL PROCEDURES

In this chapter, we concentrate on spectroscopic measurements of static and dynamic processes on fractal structures. Recall that in Sections 3.1.2.2 and 3.1.3 we dealt with light-scattering analysis of fractal systems, demonstrating how such results clarify that one and the same fractal may combine irregularities of different mass, surface, and pore fractality. Now, we discuss this and other *elastic* processes to further characterize the chemistry of fractals. In addition, we look at selected examples of chemical interest that involve dispersive spectroscopies [1]. Dispersive spectroscopies are of interest because they characterize vibrational modes that are sensitive to the connectivity of a fractal; in other words, its particular looping/branching patterns as quantified by the fracton or spectral dimension $\bar{d}$ (see Chapter 4). First, recall that various processes of transport or flux are related by one general differential equation (Section 4.1) and, second, that vibrational spectroscopy at longer wavelengths (low frequencies) probes the dynamics of longitudinal displacement modes (acoustic phonons) that are, conceptually, equivalent to translational particle diffusion. (Note further that this interrelation is independent of any fractality of the substrates.) Hence, an appropriate *time-dependent* dynamical quantity—for instance, the probability $[p(t,0)] \sim t^{-\bar{d}/2}$ of a random walker to return to a certain, previously visited site—has a *frequency-dependent* counterpart, namely, the <u>*density of vibrational states*</u> $\rho(\omega) \sim \rho^{\bar{d}-1}$ of the long-wavelength phonons (the number of vibrational eigenfrequencies per unit frequency interval): One is simply the Laplace transform of the other [2].

Recall also that anomalous diffusion (over loops, branching points, and dangling bonds) is a local phenomenon. Correspondingly, its fractons are found within narrower frequency ranges than "ordinary" vibrational phonon modes localized by the lifting of transition rules due to lack of translational symmetry of the amorphous object [3]. We encounter again the important principle of fractal properties and behavior that the size range of the probe (now a fracton frequency range) must match the size range of the object (now the connectivity domain of the fractal). Indeed, this is not different from the condition that tiling the surface of a fractal adsorbent requires adsorbates of size ranges comparable to the linear extent of the irregularities probed (Section 2.3). Therefore, fracton modes do not exist beyond the limits of too long and too short wavelengths, respectively, as such mismatch would render them blind to the details of the connectivity of the fractal substrate: Fracton modes are strictly *localized phonons*.

5.1.1   Connected Structures in Water

It is well known that liquid water is a network of hydrogen-bonded molecules in the form of random clusters; water contains regions of varying connectivity that are structured as loops and dead ends from hydrogen-bond breakage and re-formation. Fracton modes are therefore to be expected and low-frequency Raman spectroscopy is the method of choice because (a) spectral accumulation in the low-frequency region, usually below $v = \omega/2\pi \approx 50$ cm^{-1}, is readily accessible with automated, sensitive equipment, (b) sample preparation is easy (glass containers), (c) water is transparent to the optical wavelengths of exciting laser sources, (d) the apparatus does not require specialized support personnel.

Conveniently, the Raman frequency-dependent (depolarized) intensity $J(\omega)$, as recorded by the spectrometer, is modified to more useful optical quantities such as a susceptibility, $\chi''(\omega)$, or an absorption coefficient, $\sigma(\omega) \sim \omega\chi''(\omega)$:

$$J(\omega)/[1 - \exp(-\hbar\omega/kT)]^{-1} \sim \chi''(\omega) \sim \sigma(\omega)/\omega \qquad (5.1)$$

Quantity $[1 - \exp(-\hbar\omega/kT)]^{-1}$ is the sum of states, the equilibrium energy distribution of all vibrational modes of energy $\hbar\omega$ at absolute temperature T [4]. Susceptibility $\chi''(\omega)$ is related to the density of vibrational states $\rho(\omega)$ (see Glossary 2) and to radiation–phonon coupling coefficient $C(\omega)$ [5] by

$$\chi''(\omega) \sim \rho(\omega)C(\omega)/\omega \sim \sigma(\omega)/\omega \qquad (5.2)$$

Figure 5.1 presents a (natural) log–log plot of quantity $\sigma(v) \sim \rho(v)C(v)$—the effective density of vibrational states—versus linear frequency $v = \omega/2\pi$ at particular temperature of 22°C (solid curve), computed with Eq. (5.1) from depolarized low-frequency Raman data of $J(v)$ of liquid water [6, 7]. The plot also contains concordant data for a H_2O–C_2H_5OH mixture of 0.025 mol fraction C_2H_5OH (dashed line). Note that absorption coefficient $\sigma(v)$ exhibits a crossover point at $v_C = 4.1$ cm^{-1}, implying that the system is effectively homogeneous at frequencies *below* cutoff v_C because its local connectivity (loops, branching structures, and dangling bonds) is overlooked by the nonmatch with the long-wavelength scattered radiation (see Section 5.1). Hence, the observed scaling $\sigma(v) \sim v^{2.05}$ at spectral frequencies $v < 4.5$

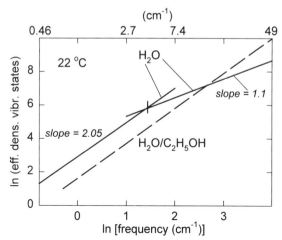

Fig. 5.1 Log–log (natural) frequency scaling plot of the effective density of states of liquid H_2O (solid lines) and of a 0.025 mol fraction C_2H_5OH in H_2O mixture (dashed line) at 22°C. The many individual data points are not shown. The crossover frequency between phonon and fracton modes, highlighted by the vertical tick, lies at $\nu_C \approx 4.1$ cm^{-1}. [Reprinted with permission from D. Majolino, F. Mallamace, P. Migliardo, F. Aliotta, N. Micali, and C. Vasi, *Phys. Rev. E* **47**, 2669 (1993), Fig. 1b, American Physical Society.

cm^{-1} ought to correspond rather closely to the density of vibrational states of *ordinary* localized phonons in Euclidian space $E = 3$ (see Glossary 2, *localized phonons*), namely, $\rho(\nu) \sim \nu^{E-1} \sim \nu^2$ [7]. On the other hand, for frequencies above crossover point ν_C, fracton modes ought to be present (see Section 5.1). Indeed, for frequencies $\nu > \nu_C$, slope s of $\sigma(\nu)$ in Fig. 5.1 has dropped from $s = 2$ to $s = 1.1$, indicating that in the expression of the density of vibrational states, $\rho(\nu) \sim \nu^{E-1}$, Euclidian dimension $E = 3$ is replaced by fracton dimension $\bar{d}$, with $\bar{d} < 2$ (see Section 4.1).

Because we do not know radiation–phonon coupling coefficient $C(\nu)$, the spectra do not allow us to deduce the numerical value of spectral dimension $\bar{d}$. However, it seems promising to explore an observed temperature dependence of crossover frequency ν_C as this may reflect changes in the water connectivities and their energies due to hydrogen-bond breakage and re-formation; Fig. 5.2 displays the appropriate data for water. Note that lowering temperatures causes a down-shift (lower frequencies) of crossover frequency ν_C, implying a *decrease* in the range of localized phonon regimes ($\nu < \nu_C$) and a corresponding *increase* in the range of (localized) fracton behavior ($\nu > \nu_C$). Now, recall that fracton behavior is found only within some linear connectivity range ξ, namely, that over which the hydrogen-bonded water system is a looped/branched fractal. Therefore, fracton–phonon crossover frequency ν_C and length scale ξ are related—reasonably by an inverse functional relationship: The larger a range ξ of the connectivity of the fractal, the lower the phonon–fracton crossover frequency

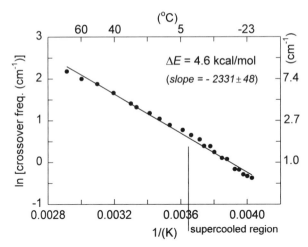

Fig. 5.2   Semi-log (natural) activation plot of crossover frequencies between phonon and fracton modes in liquid water (see also Fig. 5.1). Slope $\Delta E/k$ represents the linear least-squares calculation through the data points with its standard deviation. [Reprinted with permission from D. Majolino, F. Mallamace, P. Migliardo, F. Aliotta, N. Micali, and C. Vasi, *Phys. Rev. E* **47**, 2669 (1993), Fig. 3b, American Physical Society.]

ν_C and the wider the frequency range of the fracton regime. Indeed, we find that [6, 8]

$$\nu_C \sim \xi^{-d/\bar{d}} \tag{5.3}$$

To demonstrate the usefulness of Eq. (5.3), we assume liquid water as a hydrogen-bonded percolation cluster, with fractal dimension $d = 2.5$ ($E = 3$) for the incipient infinite cluster (see Section 3.1.1) and fracton dimension $\bar{d} = 1.33$ (Section 4.1). Thus we find $\nu_C \sim \xi^{-1.88}$ or $\xi \sim \nu_C^{-0.53}$, implying that correlation length ξ of the regions of four-bonded water molecules [9] is about proportional to the inverse of the square root of ν_C. Selecting crossover frequency ν_C from Fig. 5.2 at the extreme experimental temperatures of $-23°C$ and $70°C$, one obtains an estimate of the ratio of the system's range of connectivity at two opposite states, namely, the high-temperature (70°C) and the supercooled liquid ($-23°C$). We estimate $\xi(-23°C)/\xi(70°C) \approx 4$, an instructive and interesting result, implying that the range of four-bonded water molecule structures increases by a considerable factor upon lowering the system temperature into the supercooled regime.

Next, we consider the spectral data of the H_2O–C_2H_5OH mixture (dashed line in Fig. 5.1). Note that a fracton–phonon crossover point is no longer discernible throughout the entire frequency and temperature ranges studied [6]. Furthermore, absorption coefficient $\sigma(\nu)$ scales with ν as ν^2 throughout the entire experimental frequency range, a slope that is nearly equal to that observed for pure water under condition $\nu < \nu_C$—the localized *phonon* regime. Clearly, this shows that the scattered Raman intensities perceive the water–alcohol system as a disordered Euclidian state throughout. Ethanol, which forms three-bonded structures with water, serves as a structure breaker of the patches of four-bonded water molecules [6].

5.1.2 Silica Aggregates: Structure and Dynamics

The aggregation of solid silica particles to form sols, gels, aerosols, and aerogels is of considerable industrial and ecological-environmental importance. Particular modes of silica aggregation and its kinetics are similarly significant; we recall that agglomeration frequently generates fractal structures containing loops, dangling bonds, and other irregularities.

Depending on methods of synthesis, a variety of silica structures of wide-range chemical–physical properties are available, such as (a) preparations of silica particles in a desired range of size from aqueous media, (b) silica with surface SiOH groups able to form hydrogen bonds with water and carrying negative charges adjustable by the solution pH, (c) solution-prepared silica aggregates containing few or no SiOH surface groups upon heat treatment, and (d) formation of silica-based aerogels, a state of matter that retains an open, relatively little compacted structure. Finally, silica aggregates are chemically stable. Not surprisingly then, the agglomeration processes of primary silica particles have been followed by elastic spectroscopic methods—essentially characterizing fractal dimensions—and by inelastic spectroscopies—specifying fracton or spectral dimensions, and thereby predicting the connectivity within the fractal silica agglomerate structure. Among the former techniques, we had discussed the light scattering method in Section 3.1.2.2; the following agenda presents a more detailed account of a quantitative specification of fractal structures of silica aggregates using neutrons as radiation source.

5.1.2.1 Cationic Surfactant-Induced Flocculation of Silica Spheres

Silica particles aggregate to larger clusters by induced flocculation of their unbuffered aqueous solutions (pH 9) by cationic surfactant trimethylammonium bromide, a molecule carrying a single 10 (CH_2) chain [10, 11]. The method of small-angle neutron scattering (SANS) is particularly useful as wave vector q can be varied to probe the structure and geometry of the silica aggregates over a linear range from the diameter of a primary silica particle (200 and 360 Å), at highest settings of q, to a radius of gyration r_G (4000 Å) at its lowest settings [see Eq. (3.6) and recall that q^{-1} gives the resolution length of the method].

The outcome of the SANS experiments [10] is plotted in Fig. 5.3 in terms of (decimal) log [scattering intensity, $I(q)$] – log(wave vector, q). The following features are noticeable: (a) In the low q range, log $q < -1.8$ nm^{-1} or $q^{-1} > 600$ Å, scaling slope s decreases from $s = -0.6$ for unflocculated sol (no addition of flocculant) to $s = -2.1$ for aggregates that had been flocculated by addition of 20 mM/100 cm^3 of the cationic agent. (b) For range log $q > -1.8$ nm^{-1} or $q^{-1} < 600$ Å, slope s apparently approaches $s = -4$.

What does this tell us? First, we consider the larger linear yardstick scale of the experiments for $q^{-1} \leq 2500$ Å [12] as to greatly exceeding the diameter of a primary silica particle (200 and 360 Å). For maximal cationic concentrations (▲, Fig.

5.3), we find $s = -2.1$. Hence, the corresponding fractal dimension represents a mass fractal [see Eq. (3.8)], its dimension $d_M = -s = 2.1$ agreeing well with fractal dimension $d \approx 2.1$ predicted for reaction-limited cluster–cluster aggregation (RLA) in $E = 3$ space (see Section 3.2.2). On the other hand, agreement of $d_M = 2.1$ is poor with the mass fractal dimension $d \approx 2.5$ predicted for particle–cluster diffusion-limited aggregation (DLA, Section 3.1.2). It thus seems that silica aggregation here follows the RLA process.

But can we be fully convinced of the participation of a particular aggregation process on the sole agreement of predicted and measured dimensions? After all, fractal dimension $d_M = 2.1$ here refers to the geometrical irregularity of the silica but, *on principle*, cannot tell how the object's topology was generated. Let us then review the main steps of the preparatory method of silica aggregation [11] prior to spectroscopy and see whether they support the best-fit RLA mechanism. First, close approach of silica particles is favored upon neutralization of their charges by the action of the $N(CH_3)_3$ group of the surfactant molecules. Second, stable adherence between adjacent silica particles is promoted through the hydrophobic interactions of the CH_2 chain segments of the cationic surfactant. Therefore, this two-step flocculation process strongly hints at the intervention of a simple particle–cluster DLA process of instant and stable adherence upon encounters—rather than the slower RLA (reaction-limited) mechanisms via multiple interparticle collision attempts

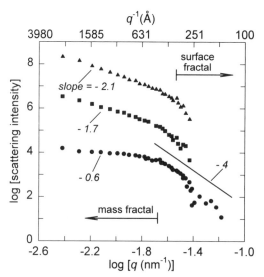

Fig. 5.3 Decimal log–log scaling plot of SANS scattering intensity data for polydispersed silica aggregates obtained through flocculation by a cationic surfactant. Concentrations of surfactant in mM/100 g: 20 (▲), 4 (■), 0 (●). The individual plots are shifted vertically for better viewing. [Reprinted with permission from K. Wong, B. Cabane, R. Duplessix, and P. Somasundaran, *Langmuir* **5**, 1346 (1989), Fig. 3, Copyright 1989 American Chemical Society.]

prior to reactive aggregation (see Section 3.2.2). Nevertheless, as we just discussed, the measured fractal dimension $d = 2.1$ does not fit a DLA process but agrees with that predicted for the RLA mechanism.

A way out of this dilemma is found by realizing that aggregation of the primary silica particles leads to a considerably wide distribution of cluster masses in the gel-like aggregate—the global percolation cluster exists in the presence of many smaller clusters of varying sizes. Recall that several types of cluster size distributions and their largest radii were briefly mentioned during the discussion of size-exclusion chromatography (Section 3.1.1); the same general concepts are now used in the necessary mass averaging of the light-scattering process to account for a distribution of aggregate size. It turns out that this modifies mass fractal scattering law $I(q) \sim q^{-d}$ [Eq. (3.8)] of narrow cluster-size distributions to

$$I(q) \sim q^{-d(3-\tau)} \tag{5.4}$$

of wide cluster-size distributions under a *polydispersity exponent* τ. (Greek letter τ is used in the literature on polydispersity and should not cause confusion with other uses of τ, such as a time interval.) Using numerical value $\tau \approx 2.2$ (see Glossary 2), we find that mass fractal dimension $d = 2.5$ of particle–cluster DLA for narrowly mass-distributed percolation clusters in $E = 3$ is reduced to mass fractal dimension $d \approx 2$ of particle–cluster DLA for polydisperse samples. This value is now in much better agreement with SANS result $d = 2.1$ (Fig. 5.3) and retains the very reasonable hypothesis that silica aggregation follows the DLA mechanism after all.

Undoubtedly, this outcome is a warning that we must not blindly assign slopes of scaling plots of elastic scattering data to fractal dimensions from computer-generated predictions, however closely their results agree with the experiment.

We now discuss the observed $I(q) - q$ scattering data in Fig. 5.3 for larger q (smaller q^{-1}), hence resolution lengths of the order of the diameter of the primary silica particle, about 250 Å and smaller. To help the eye, slope $s = -4$—corresponding to surface fractal dimension $d_S = s + 6 = 2$ [Eq. (3.9)]—is drawn into the figure. Two observations are notable: The primary silica particles of the aggregates have classical surfaces, and particle diameter 250 Å is the lower cutoff of the validity of the d_S data because smaller diameters imply surface dimensions of $d_S < 2$.

Finally, some comments on the data under low or zero flocculent concentrations exhibited in Fig. 5.3. The latter case is easy to appreciate: In the absence of flocculent aggregation is relatively minimal and the slope of the scaling plot is correspondingly constant under wave vector q once its resolution significantly exceeds the size of the primary silica particles [13]. On the other hand, the data at 4 mM/100 cm^3 flocculant are difficult to understand as the concentration level is deemed sufficient to neutralize the active SiOH sites on the primary particles and the mechanism is taken to be independent of the CH_2 chain length [10].

5.1.2.2 *Fractons in Silica Aerogels*
What constitutes a silica aerogel? An aerogel has the liquid phase of its originating gel structure replaced by a gaseous phase; shrinkage during the drying process is then

avoided [11]. Silica aerogels are relatively noncompacted structures of rather low densities and are obtained here by hydrolysis of tetramethoxysilane with a 0.05 N aqueous ammonia solution in the presence of methanol—a coagulation process [11]—and subsequent oxidative heat treatment at 500°C [14]. Such a synthetic procedure leads to fairly monodisperse colloidal aerogels at densities of 73–142 kg/m³, with mean diameter of primary particles of 32 ± 12 Å.

Figure 5.4 displays a (decimal) log–log plot of small-angle neutron scattering (SANS) results of a series of such silica aerogels. The observed slopes $s = -1.7$ for reciprocal wave vector $14 < q^{-1} < 50$ Å relate to dimension $d \approx 1.7$, implying that the aerogels probed within this resolution range are mass fractals [Eq. (3.8)]. Dimension $d = 1.7$ agrees, formally, with predictions from diffusion-limited cluster–cluster aggregation in $E = 3$. It also agrees, approximately, with fractal dimension $d = 1.5–1.7$ of silica cluster–cluster aggregation analyzed by image analysis—refer particularly to Fig. 3.2. Consequently, aggregation of primary silica particles to an aerogel by hydrolysis and oxidative heat treatment seems to arise from cluster–cluster DLA, whereas surfactant-induced flocculation of primary silica particles follows particle–cluster DLA (see Section 5.1.2.1).

Within regime $q^{-1} < 14$ Å, Fig. 5.4 indicates a surface fractal dimension of $d_S = 2$ deduced from slope $s = d_S - 6 = -4$ [see Eq. (3.9)], a value agreeing with the corresponding surface dimension from a silica surface tiling analysis (Section 2.3.3). It also agrees with surface dimension d_S of flocculated silica although the primary aerogel particles are about a factor of 10 smaller than those making up the flocculated aggregates (Section 5.1.2.1).

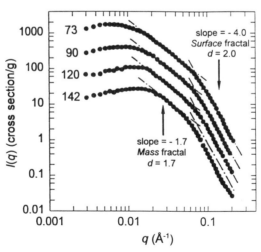

Fig. 5.4 Decimal log–log scaling plot of SANS scattering intensity data for four mutually self-similar silica aerogels at indicated densities (in kg/m³). The individual plots are shifted vertically for better viewing. [Reproduced with kind permission from E. Anglaret, A. Hasmy, E. Courtens, J. Pelous, and R. Vacher, *J. Non-Cryst. Solids* **186,** 131 (1995), Fig. 1, Elsevier Science-NL, Sara Burgerhartstraat 25, 1055 KV Amsterdam, The Netherlands.]

Of course, an identical mass fractal dimension does not predict identical aspects of local connectivity within the aggregates. To establish the latter, we need to obtain spectral dimension $\bar{d}$ of the fractal aerogel substrates of Fig. 5.4, which is most directly (although *not* most conveniently) accomplished with inelastic neutron scattering techniques because they yield the density of vibrational states, $\rho(\omega) \sim \omega^{\bar{d}-1}$, without having to bother about phonon–photon coupling coefficient $C(\omega)$ [see Eq. (5.2)]. The dashed line in Fig. 5.5 displays a (shortened) range of the resulting data, taken at 60K, in terms of the logarithms of the density of vibrational states (multiplied by absolute temperature T) and of energy $\hbar\omega$, with $\bar{d} = 1.3 \pm 0.1$ [14, 15], a value close to the fracton dimension of DLA or percolation clusters (see Section 4.1).

The shaded region in Fig. 5.5 delineates a fracton mode $\Rightarrow$ particle mode crossover to the higher frequency region of particle modes [16]. Note that this crossover brackets the *higher* frequency range of the fracton region; its low-frequency limit, namely, the fracton–phonon crossover, is at even lower frequencies (0.03 cm^{-1}) than can be accommodated in Fig. 5.5 [14, 15].

Figure 5.6 shows results from *Brillouin scattering* of the aerogels, a technique most suitable to probe the very low-frequency phonon–fracton crossover regions established for the open-structured aerogels, as just mentioned. Brillouing theory yields frequency–wave vector dispersion relation

$$\omega(q) \sim q^{d/\bar{d}} \tag{5.5}$$

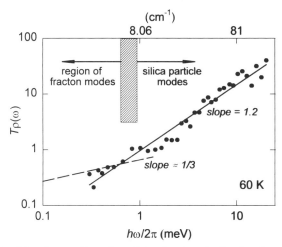

Fig. 5.5 Decimal log–log scaling plot of the density of vibrational states from inelastic neutron scattering of a silica aerogel (of density 185 kg/m^3) at 60 K about the fracton–particle mode crossover frequency (shaded area). The data have been multiplied by absolute temperature T. [Reproduced with permission from R. Vacher, T. Woignier, J. Pelous, G. Coddens, and E. Courtens, *Europhys. Lett.* **8**, 161 (1989), Fig. 2, Europhysics Letters.]

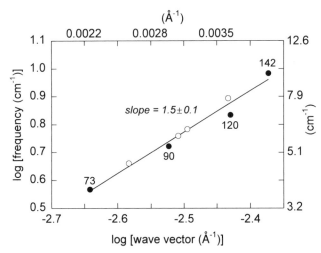

Fig. 5.6 Log–log (decimal) plots of fracton dispersion $\omega(q)$ from Brillouin scattering for a series of mutually self-similar silica aerogels. Numbers at the solid data points refer to the densities (kg/m^3) shown in the SANS data of Fig. 5.4; open circles pertain to silica aerogels of intermediate densities. [Reproduced with kind permission from E. Anglaret, A. Hasmy, E. Courtens, J. Pelous, and R. Vacher, *J. Non-Cryst. Solids* **186**, 131 (1995), Fig. 4, Elsevier Science-NL, Sara Burgerhartstraat 25, 1055 KV Amsterdam, The Netherlands.]

a formulation readily understood by realizing that Eq. (5.5) is simply Eq. (5.3) suitably modified by relating scaling range ξ of Eq. (5.3) to inverse wave vector or resolution length q^{-1} (see Section 3.1.2.2). Hence, $\xi \sim q^{-1}$. Then, by using mass fractal dimension $d = 1.7$ (Fig. 5.4) and dispersion slope $s = d/\bar{d} = 1.5$ (Fig. 5.6) leads to silica aerogel fracton dimension $\bar{d} = 1.7/1.5 = 1.1 \pm 0.1$ [14].

In summary, we obtained spectral dimension $1.1 \leq \bar{d} \leq 1.3$ of (mutually self-similar) silica aerogels of densities $73 - 142$ kg/m^3 by inelastic neutron and Brillouin scattering with experimental uncertainty of ± 0.1. The middle value within this range, $\bar{d} \approx 1.2$, agrees reasonably well with predictions from cluster–cluster aggregation in $E = 3$ under negligible formation of loop structures [17].

5.1.3 Electron-Transfer Reaction of the Adsorbed Anthracene Radical Cation

The last experimental section of Chapter 5 deals with spectroscopic laboratory results [18] of excitation transport by anomalous diffusion within the diffusion-limited scheme $A + B \Rightarrow 0$, a reaction that had occupied us in Chapter 4. In brief, anthracene was preadsorbed on silica gel, the anthracene radical cation ($A^{\cdot +}$ for short) generated in small concentrations by the action of a 355-nm laser, and electron transfer to the activated anthracene adsorbate from coadsorbed electron donor N,N,N',N'-tetramethyl-1,4-phenylenediamine (TMPD) recorded at delay times within the range of 0.05–9.17 ms by the relative changes of the anthracene reflectance band at 715 nm (time-resolved transient difference spectroscopy).

The underlying diffusion-limited kinetic scheme $A^{\cdot +} + TMPD \Rightarrow A + TMPD^{\cdot +}$ is a pseudo-first-order reaction in $A^{\cdot +}$ because the concentration of TMPD remains essentially constant, there is no back-reaction, and the electron-donating step is very fast. Integration of its classical rate expression $-d[A^{\cdot +}]/dt = K_0[A^{\cdot +}]$ therefore gives

$$[A^{\cdot +}] = [A^{\cdot +}]_0 \exp(-K_0 t) \qquad (5.6)$$

with [TMPD] taken into rate constant K_0 and subscript 0 indicating again initial values.

Admitting the possibility of anomalous diffusion, we replace rate constant K_0 by rate coefficient $K_0 t^{-\beta}$ according to Eq. (4.3), which modifies Eq. (5.6) into the stretched exponential form [19]

$$[A^{\cdot +}]/[A^{\cdot +}]_0 = \exp(-K_0 t^{1-\beta}) \qquad (5.7)$$

Taking natural logarithms of Eq. (5.7) and raising the resulting relation by power $1/(1-\beta)$ leads to

$$(\ln\{[A^{\cdot +}]_0/[A^{\cdot +}]\})^{1/(1-\beta)} \sim t \qquad (5.8)$$

Consequently, if the diffusive electron-transfer reaction on the silicon fractal substrate is anomalous ($\beta \neq 0$), a plot of $\ln([A^{\cdot +}]_0/[A^{\cdot +}])$ versus t should *diverge* from linearity. On the other hand, assuming that the excitation diffuses over a DLA or percolation cluster under spectral dimension $\bar{d} \approx 4/3$, we find $\beta = 1 - \bar{d}/2 = 1/3$ [Eq. (4.10b)] and set the exponent in Eq. (5.8) accordingly to $3/2$.

The respective outcomes are shown in Fig. 5.7. Note that the upper plot ($\beta = 0$) is not linear throughout, whereas the lower plot gives a satisfactory straight line through the entire time domain. Consequently, diffusional excitation transport is anomalous with $\bar{d} \approx 4/3$.

Undoubtedly, these experimental results and evaluations in terms of a fracton picture can be taken as a fine example of the relevance of the theoretical rate expressions of diffusion-limited deactivation of a large adsorbed radical cation under anomalous diffusion. Further, it seems rather certain that the pathway of the deactivation process—at least over the first 10 ms of its excitation decay—probes the connectivity of a fractal structure apparently closely related or equal to a percolation or DLA cluster (principally its branching pattern).

5.2   A SUMMARY

Various spectroscopic methods of several fractal aspects have been discussed in this chapter, allowing us to draw some conclusions on their relative merits. Low-frequency Raman spectroscopy and required sample handling are widely accessible and relatively easy to perform. Scattered intensity can be collected rather close to the frequency of the exciting laser line by modern instrumentation (see Section

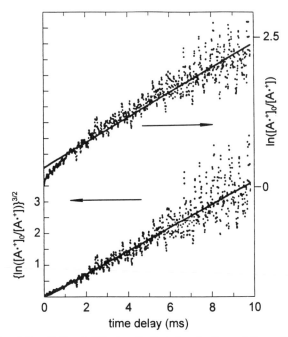

Fig. 5.7 Graph of Eq. (5.8) of diffusion-limited deactivation of the anthracene radical cation by electron donation from N,N,N',N'-tetramethyl-1,4-phenylenediamine, both adsorbed on silica gel. The upper graph, arbitrarily shifted for better viewing, displays regular diffusion ($\beta = 0$). The lower graph shows anomalous diffusion under fracton dimension $\bar{d} = \frac{4}{3}$ ($\beta = \frac{1}{3}$). [Reprinted with permission from D. R. Worrall, S. L. Williams, and F. Wilkinson, *J. Phys. Chem. B* **101,** 4709 (1997), Fig. 8, Copyright 1997 American Chemical Society.]

5.1.1) and data averaging, by scan accumulation, is routine. Theory is straightforward and well-known except for photon–phonon coupling coefficient $C(\omega)$ [20]. Low-frequency Raman spectroscopy may also be used to explore other fractal/fracton aspects of irregular systems, such as the local structure of aggregating solvent–solute systems around the percolation limit, by appropriate variations of concentration and temperature.

Application of inelastic neutron scattering spectroscopy to obtain spectral dimension $\bar{d}$ from the density of vibrational states has been somewhat disappointing. In spite of the transparency of the theoretical approach, data evaluation seems rather convoluted [15]—assuredly to a chemist curious about using the method. In addition, the experimental precision seems rather low, hardly warranting the lengthy experimental manipulations [21] and the required stay away from one's laboratory—however pleasant the other location (Grenoble, France). Apart from this, you need dedicated support staff at the institution to help you through.

As we have seen, Brillouin scattering is a useful and supporting technique for obtaining frequency–wave vector dispersion relations around *very* low-frequency phonon–fracton crossover regions—those of relevance to tenuous and open-structured fractals. It is doubtful that many chemists are presently familiar with this technique or even aware of it.

Regarding studies of excitation transfer on the basis of time-resolved transient spectroscopy, we note that results are readily applicable to theoretical models—given the usually simple reaction schemes involved in highly diluted, low-concentration active species.

With this, the chapters dealing with *fractons* are concluded and we return to purely structural fractal aspects. One piece of advice: Chemists interested in applying fracton concepts ought to glance now and then at issues of *Physical Review Letters*—of all journals, indeed. Among articles on strange new particles, cosmological constants, and other esoterica (dear to physicists), it is usually there where you will find short, relevant, precise, and informative reports on newest developments in fracton theory [22].

NOTES AND REFERENCES

[1] In the elastic process, the frequency of the incident radiation is equal to that of the scattered radiation and no energy is exchanged with the internal degrees of freedom of the substrate. Dispersive spectroscopy displays the change of a measured optical property with frequency under exchange of energy between the incident radiation and energy levels of the irradiated system.

[2] The chemist using a Fourier transform infrared (FTIR) spectrometer, now in common use in many laboratories, will realize that the interferogram scanned over a distance measure (the motion of the mirror) is converted into a spectrum over a frequency measure by a Fourier transform.

[3] K. D. Möller and W. G. Rothschild, *Far-Infrared Spectroscopy* (Wiley, New York, 1971), pp. 516–522.

[4] W. G. Rothschild, *Dynamics of Molecular Liquids* (Wiley, New York, 1984); see Appendix K. Quantity $J(\omega) \sim \chi''(\omega)/[1 - \exp(-\hbar\omega/kT)]^{-1}$ is the initial-state-averaged transition probability; see H. B. Levine and G. Birnbaum, *Phys. Rev.* **154**, 86 (1967). For $\hbar\omega/kT \ll 1$, we get $\sigma(\omega) \approx \omega^2 J(\omega)$ [see Eq. (5.1)].

[5] R. Shuker and R. W. Gammon, Raman-scattering selection-rule breaking and the density of states in amorphous materials, *Phys. Rev. Lett.* **25**, 222 (1970); V. Mazzacurati, M. Nardone, and G. Signorelli, Depolarized Rayleigh scattering in 10 M amorphous and liquid KOH aqueous solutions, *J. Chem. Phys.* **66**, 5380 (1977); E. Duval, N. Garcia, A. Boukenter, and J. Serughetti, Correlation effects on Raman scattering from low-energy vibrational modes in fractal and disordered systems. I. Theory, *J. Chem. Phys.* **99**, 2040 (1993).

[6] D. Majolino, F. Mallamace, P. Migliardo, F. Aliotta, N. Micali, and C. Vasi, Spectral evidence of connected structures in liquid water: Effective Raman density of vibrational states, *Phys. Rev. E* **47**, 2669 (1993).

[7] Classical solid-state theory predicts $\rho(\omega) \sim \omega^2$ in $E = 3$ for the low-frequency density of states in range $0 < \omega < \omega_D$, where ω_D is the Debye frequency.

[8] For a more general derivation, see T. Nakayama, K. Yakubo, and R. L. Orbach, Dynamical properties of fractal networks: Scaling, numerical simulations, and physical realizations, *Rev. Mod. Phys.* **66**, 381 (1994), p. 394. Equation (5.3) can be rationalized by a simple argument: We relate the cumulative number of fracton modes under spectral dimension $\bar{d}$, $\mathcal{N}(\omega) \sim \int_0^\omega \rho(\omega)d\omega \sim \omega^{\bar{d}}$ (Glossary 2, *density of vibrational states*), with the number of self-similar, connected elements over correlation length ξ under fractal dimension d, $\mathcal{N}(\xi) \sim \xi^{-d}$ [Eq. (1.1)]: Hence we get $\omega^{\bar{d}} \sim \xi^{-d}$.

[9] H. E. Stanley and J. Teixeira, Interpretation of the unusual behavior of H_2O and D_2O at low temperatures: Tests of a percolation model, *J. Chem. Phys.* **73**, 3404 (1980).

[10] K. Wong, B. Cabane, R. Duplessix, and P. Somasundaran, Aggregation of Silica Using Cationic Surfactant: A Neutron-Scattering Study, *Langmuir* **5**, 1346 (1989). The report does not indicate the units of wave vector q, but it is clear from the general results and Fig. 1 in [10] that q is in units of reciprocal nanometer, $nm^{-1} = 0.1\ \text{Å}^{-1}$.

[11] R. K. Iler, *The chemistry of silica* (Wiley, New York, 1979). Silica initially aggregated in aqueous media and then heated to high temperatures (500°C) possesses a partial surface cover of ionizable SiOH groups. Surfactant-induced *flocculation* between primary silica particles, a gel formation process, takes place by neutralization of their negative charge by the positive head of the surfactant molecule and subsequent interaction of the long CH_2 chains between surfactant molecules affixed to neighboring silica particles (micelle formation). *Coagulation* between silica particles proceeds by metal cation-induced loss of silanol sites needed for silica–water hydrogen bonding (dehydration).

[12] From $q^{-1} = 1/10^{-2.4} = 251$ nm or 2510 Å.

[13] The q-independent scatte.ing intensity for the sol phase of unflocculated primary silica particles (lowest curve in Fig. 5.3) gives the radius of gyration of the independent, individual silica particles [see Eq. (3.7)].

[14] E. Anglaret, A. Hasmy, E. Courtens, J. Pelous, and R. Vacher, Fracton dimension of a mutually self-similar series of base-catalyzed aerogels, *J. Non-Cryst. Solids* **186**, 131 (1995).

[15] R. Vacher, T. Woignier, J. Pelous, G. Coddens, and E. Courtens, The Density of Vibrational States of Silica Aerogels, *Europhys. Lett.* **8**, 161 (1989). Neutron spectroscopic work frequently scales the independent variable in terms of energy (meV = millielectron volt) instead of energy (cm^{-1}). Here 1 meV is equivalent to 8.06 cm^{-1}.

[16] Particle modes involve the usual stretch/bend/deformation displacement coordinates of segments of larger masses and/or weaker force constants within materials; their spectral bands appear near 600 cm^{-1} and range to frequencies in the far-infrared. Particle modes are not of our concern here except to demonstrate that they determine the upper frequency limit of the fracton region.

[17] For essentially loopless structures such as DLA and percolation clusters, theory predicts $\bar{d} = 2d_L/(d_L + 1)$, where d_L is the dimension of the chemical distance L—the shortest path *on* the cluster between two of its points—under mass–distance scaling $M(L) \sim L^{d_L}$. With $d_L = 1.42$, we find $\bar{d} = 1.17$ for cluster–cluster aggregation in $E = 3$. See Eq. (4.6) and Table 6 of [3] in Section 4.1.

[18] D. R. Worrall, S. L. Williams, and F. Wilkinson, Electron transfer reactions of anthracene adsorbed on silica gel, *J. Phys. Chem. B* **101**, 4709 (1997).

[19] Rate-determining concentration fluctuations or inhomogeneous collisional encounter frequencies play no role in this sytem (see Eq. (4.12), and [26] in Section 4.2.4).

[20] A. P. Sokolov, U. Buchenau, W. Steffen, B. Frick, and A. Wischnewski, Comparison of Raman- and neutron-scattering data for glass-forming systems, *Phys. Rev. B* **52,** R9815 (1995).

[21] A. J. Dianoux, Neutron scattering by low-energy excitations of disordered solids, *Philos. Mag. B* **59,** 17 (1989).

[22] See (a) [10] in Section 4.2, (b) D. Cassi and S. Regina, Random Walks on Bundled Structures, *Phys. Rev. Lett.* **76,** 2914 (1995), and (c) R. Burioni and D. Cassi, Universal properties of spectral dimension, *Phys. Rev. Lett.* **76,** 1091 (1995).

6

CHEMICAL DEGRADATION

6.1 INTRODUCTION: FRACTAL ASPECTS IN MATERIAL DETERIORATION PROCESSES

Appearance of fractal structures in deterioration processes of materials of all types and usage by chemists is quite obvious and frequent, once one knows how and where to look for them. For instance, crack lines of solid materials or the tear line of paper are irregular curves, often exhibiting self-similarity or self-affinity of varying degrees over wide size ranges of the object. Intimately connected with such patterns are the processes leading to or initiating them, such as surface corrosive dissolution. Their fractal aspects introduce model concepts of wide applicability to phenomena that seem to have nothing in common with degradation or deterioration of materials. Indeed, fractal aspects within the various sections of this chapter, from surface smoothing by ablation to its opposite, surface roughening by corrosion, will be seen to present two sides of the same coin. Thus they furnish yet another reminder of the generality and usefulness of fractals in chemistry.

6.1.1 Ablation: Ruggedness Changes in Food Particles

Who has not opened a fresh box or carton of some powdered foodstuff or other household supply and not thought "Well, it is less than I paid for!" However, once the supply has dwindled and the bottom of the box is reached, a residue of accumulated fine powder is quite noticeable or—in the other extreme—large bits and pieces of its originally freely flowing content. The explanation is obvious: Particle ablation and particle compaction. Many substances of daily use are agglomerates of their smaller constituent particles, therefore less dense than mass-Euclidian materials and of lower mechanical strength. Hence, such materials are often exceptionally

116

sensitive to unavoidable frictional events between adjacent particles during transport and daily use.

Instant coffee is such agglomerated powder and "ablation" that it *lowers* the surface fractality of the agglomerate by removal of preexistent surface irregularities via frictional events, an example I unexpectedly found by random walk along the journal shelves in the Wayne State University Science Library. Clearly, particle ablation may lead to undesirable effects other than purely esthetical; for instance, the resulting increased compaction (the diminution of interparticle voids) may lead to an increased density, a decreased solubility, or slower rate of dissolution, or simply a loss of surface area. Long-standing quality control methods are based on determinations of instant coffee bulk densities under varying product heights in graduated cylinders, subjected to a lengthening duration of mechanically administered taps. Various model concepts, some using four-parameter fitting, have long since been employed to analyze the data and to compress the information into a useful equation of predictive capability. Lately, a non-classical approach to analyze the treatment-induced smoothing of the instant coffee particles was tried by measuring the *decrease* of the fractal particle coastline dimension relative to fresh material [1].

Such fractal approach has several advantages over current methods. First, the deterioration of the material by ablation can be characterized by a *single* number, a fractal perimeter dimension. Thereby, the use of complicated models with several input parameters can be avoided. Second, the fractal method is more sensitive than the bulk-pressure analysis mentioned above because it is the very perimeter of the particle where its deterioration occurs. Third, fractal dimensional analysis can be performed readily, for instance, with the well-tried method of "divider stepping" [1] around the perimeter of the aggregates (see Section 3.1.2.1.)

Figure 6.1 depicts a histogram of the perimeter dimension of instant coffee of 16/20 mesh size (average particle size 1.02 mm) as a function of the number of mechanical taps applied to its container. Note that the perimeter of the particles represents boundary fractals, of dimension $d_p = 1.11$ for a fresh sample. Although the decrease of the perimeter dimension d_p during increased taps is not large on an absolute scale, the drop of d_p from 1.11 to $d_p = 1.07$ under 10^4 taps conforms well with the visual difference between fresh and treated instant coffee particles: The human eye is very perceptive of small topological differences in smaller objects and can well discern, as is demonstrated in Fig. 6.1, the silhouettes of a fresh (upper left, $d_p = 1.11$) and a treated (upper right, $d_p = 1.07$ after 10^4 taps) instant coffee particle.

It seems particularly worthwhile to explore the time development of the ablation process—the accumulated damage from continual particle ablation—by scaling the logarithm of the fractal perimeter dimensions with the logarithm of the number of taps applied over the measured time interval. Perhaps experimental precision is wanting but the result, $d_p \sim t^{-\gamma}$ with $\gamma = 0.0046 \pm 0.0007$ displayed in Fig. 6.2, is nevertheless quite instructive. It implies that most of the damage occurs during the initial stages of the ablation process because the inferred power law predicts that the ablation rate, $-(d/dt)d_p \sim \gamma t^{-(1+\gamma)}$, slows down with increased duration of

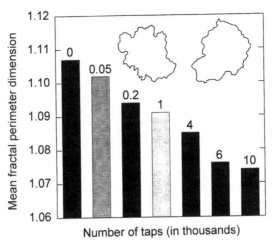

Fig. 6.1 Histogram of coastline (perimeter) fractal dimension d_p of instant coffee particles as function of the duration of an ablation process by mechanical tapping. The silhouettes of a typical instant coffee aggregate before (left-upper inset) and after 10^4 taps (right) are also shown. [Reprinted with permission from B. J. Barletta and G. V. Barbosa-Cánovas, *J. Food Sci.* **58**, 1030 (1993), Figs. 3a, 5a, 5c, Institute of Food Technologies.]

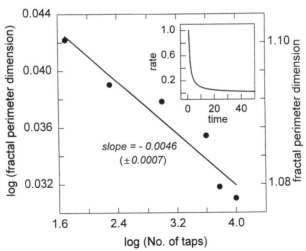

Fig. 6.2 Decimal log–log plot of perimeter dimensions d_p and number of mechanical taps of the data of Fig. 6.1. The value of the linear least-squares slope is -4.6×10^{-3}, its standard deviation amounts to $\pm 0.7 \times 10^{-3}$. *Inset:* Rate of ablation with time (number of taps) by the decrease of perimeter dimension, d_p.

attrition. Indeed, it would require of the order of 10^{10} taps to reach a reasonably smooth perimeter of, say $d_p = 1.01$. The inset of Fig. 6.2 gives the general idea.

Considering unavoidable initial handling of such materials, this outcome is not good news to the manufacturer and distributor.

6.1.2   Surface Roughening by Dissolution

Gradual corrosion of materials by various ambient agents is an ever-present phenomenon. The process usually begins and continues at the outside of a material, its surface, where pitting and other surface-roughening and surface-enlarging steps take place. We expect that their dynamics contain random aspects—recall the surface pitting model in Section 1.2—but they turn out to be not entirely random: A bias is apparent in the process in the sense that an event retains a (partial) memory of the system state at the preceding instant. Therefore, to correctly describe the observations, the randomness of the process has to be tampered with.

The underlying theory is based on principles of the so-called *fractional Brownian motion* [2], an extension of the normal Brownian walk of a completely random outcome of events in terms of root-mean-squared distance traveled during time t, $\langle R^2 \rangle^{1/2} \sim t^{1/2}$ (see Section 4.1). Briefly described, the difference between the values of random variable $X(t)$ at two succeeding times or positions, $X(t) - X(t_0)$, $t_0 < t$, is now written in the form (for long times) of

$$X(t) - X(t_0) = \eta_G |t - t_0|^H \equiv B_H(t) \quad 0 < H \leq 1 \qquad (6.1a)$$

where $B_H(t) \sim t^H$ expresses an extended diffusion model that includes regular Brownian random walk if $H = \frac{1}{2}$ but, as we shall see, allows for biased random walk scenarios depending on the value of exponent $H \neq \frac{1}{2}$, sometimes named the "Hurst" exponent. (Gaussian-distributed random variable η_G is usually omitted; it is only of concern for an actual modeling procedure [2].) To see Eq. (6.1a) from a different angle, we write the variance of $X(t)$,

$$\langle [X(t) - X(t_0)]^2 \rangle \sim |t - t_0|^{2H} \qquad (6.1b)$$

which, for $H = \frac{1}{2}$, is equivalent to Einstein's relation $\xi^2 \sim \langle R^2 \rangle \sim K_0 t$ of regular Brownian walk (see above and Section 4.1). We now have to investigate what it means to have allowed $H \neq \frac{1}{2}$.

Foremost, Eq. (6.1a) under $H \neq \frac{1}{2}$ admits that the "dice are loaded," so to speak. For $H < \frac{1}{2}$, variable $X(t)$ reflects, with some probability, a *trend* in the opposite sense to that of its preceding characteristics: $B_H(H < \frac{1}{2})$ is said to show "antipersistent" behavior. For $H > \frac{1}{2}$, a certain trend in the manifestations of random variable $X(t)$ is maintained during the following events: $B_H(H > \frac{1}{2})$ exhibits persistent behavior.

Note that B_H is a self-affine mapping; independent and dependent coordinates are scaled by different factors: $b^{-H} B_H(bt) = B_H(t)$ (see *scaling*, Glossary 1, as well as Note [7] in Section 1.2 and Fig. 1.4).

To get a better understanding of the following data—electrodissolution of the (111) surface of Ag metal by aqueous perchloric acid [3]—we assume that fractional Brownian motion is a useful descriptor of essential aspects of the diffusion-limited reactions resulting in surface corrosion or surface roughness. The process is certainly random but perhaps not completely random due to some specific metal–acid interactions. But first we discuss the experimental method and evaluation of the data.

Roughening of the originally smooth surface with lengthening reaction times is observed in situ by imaging scanning tunneling microscopy (STM) [4]. The stylus is usually moved along a straight path over the surface; it therefore probes the irregularities of the *profile* (or contour) of the rough surface (see Fig. 1.5). Numerical evaluation uses the following definitions and parameters: A profile of a flat surface of N points and length L, drawn along the x axis, exists at time $t = 0$. Quantity $h(x,t)$, in a direction normal to L, is considered the space–time random variable describing the instantaneous height of the corrosion-induced surface irregularities. Its standard deviation (averaged over spatial coordinates x)

$$w(L,t) = \{(1/N)\sum_x(\langle[h(x,t)]^2\rangle - \langle h(x,t)\rangle^2)\}^{1/2} \qquad (6.2)$$

is a measure of width $w(L,t)$ of the surface irregularities recorded at time point t during the corrosion process and at linear extent L of the corroded surface.

Figure 6.3 highlights the methodology, displaying height $h(x)$ at positions x from model computations of $B_H(t)$ [Eq. (6.1a)] with $H = 0.30$ (antipersistent), $H = 0.5$

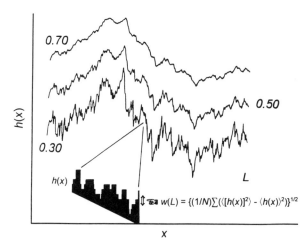

Fig. 6.3 Computer-modeled profiles of irregular surface features by (top to bottom) persistent, completely random, and antipersistent fractional Brownian motion. Irregularity height $h(x)$ is normal to the surface and sampled along direction x over total length L, with zooming-in on its root-mean-square width $w(L)$ over a small segment of profile $H = 0.3$. [Reprinted with permission from J. Krim and J. O. Indekeu, *Phys. Rev. E* **48**, 1576 (1993), Fig. 1, American Physical Society.]

(Brownian walk), and $H = 0.7$ (persistent) surface irregularity [5]. Also shown is an explanatory zoomed-in sketch of width $w(L)$ of the surface of linear extent L after surface irregularity growth has reached a steady state. Figure 6.4 shows actual STM surface profiles of developing surface instabilities, sampled at six succeeding times between 0 and 917s, from imaging a 85×85-nm^2 silver surface over basis length $L \approx 38$ nm [3].

We evaluate the profile width data by dynamic scaling [6]. Figure 6.5 displays the time power law growth of root-mean-square STM width w_{STM} [Eq. (6.2)],

$$w_{STM} \sim t^{\beta} \tag{6.3a}$$

of the metal surface under current density 30 μA/cm^2. Note that the slope is linear under *dynamic* exponent $\beta = 0.36$ up to $\tau \approx 600$ s, thereby implying nonequilibrium growth of the irregular solid-liquid interface until saturation is reached at 600 s. (Recall that we had encountered growth–time power law patterns previously, namely, during diffusion-limited electrodeposition on silver, Section 3.2.1.1)

Figure 6.5 also shows that electrodissolution run under the low current density 4 μA/cm^2 does not generate significant surface irregularities. Although the process removes metal layer-by-layer at about 150 Ag-monolayers in 10^4 s, the initial surface smoothness is retained [3]. Fractal aspects are not apparent at such low Faradic effects; the changes at the Ag surface occur under equilibrium conditions throughout.

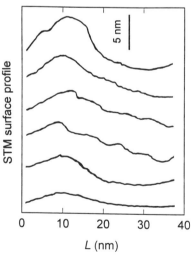

Fig. 6.4 Scanning tunneling microscopy sequential images of the profile of a (111) silver surface with increasing times of electrodissolution in HClO$_4$ of 0, 151, 355, 418, 625, and 917 s (bottom to top) at 30-μA/cm^2 current density. The absolute scale is given by the vertical bar. [Reprinted with permission from M. E. Vela, G. Andreasen, R. G. Salvarezza, A. Hernández-Creus, and A. J. Arvia, *Phys. Rev. B* **53**, 10 217 (1996), Fig. 5, American Physical Society.]

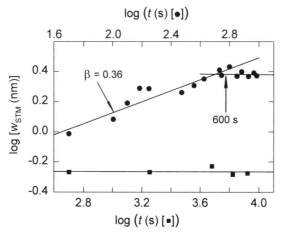

Fig. 6.5 Decimal log–log scaling plot of the time dependence of the growth of irregularity width $w \sim t^\beta$ on a (111) silver surface in 1 M $HClO_4$ under current density 4 $\mu A/cm^2$ (■) for prolonged times, and under 30 $\mu A/cm^2$ (●) for shorter times. Temperature 298 K. [Reprinted with permission from M. E. Vela, G. Andreasen, R. G. Salvarezza, A. Hernández-Creus, and A. J. Arvia, *Phys. Rev. B* **53**, 10 217 (1996), Fig. 6, American Physical Society.]

Figure 6.6 demonstrates scaling of w_{STM} with distance L over the basis length of the (111) Ag surface, following attainment of steady-state conditions ($t > \tau$). We again observe power law characteristics of w_{STM} [6].

$$w_{STM} \sim L^\alpha \tag{6.3b}$$

with $\alpha = 0.86$ for $\log L < 1.5$. For $L > 32$ nm, the values of α are seen to level off (crossover): Surface roughness width w_{STM} saturates with surface base length L. Parameter α, the static or *roughness exponent*, therefore describes the shape of the self-affine interface at steady-state conditions.

Now, we equate exponent α of Eq. (6.3b) with exponent H of fractional Brownian motion [7] [see Eq. (6.1a)]. The relation is not precise; we shall see later that different length scales may be involved with α and H. However, exponent H gives a very useful and quick overview of bias within the random sequence of the corroding events. For instance, the results of Fig. 6.6 imply strongly persistent behavior, $\alpha \approx H = 0.86$. The corroded Ag surface is relatively smooth—$\alpha = H = 1$ denoting complete absence of surface roughness—because each sequential step within the rate-determining diffusive motions retains a partial *positive* memory of the preceding step (see Fig. 6.3), on average.

The numerical agreement between the widely diverse phenomena involving corrosion-induced self-affine interfaces on one hand and electrodeposited self-affine interfaces on the other is indeed remarkable, both extreme mechanisms apparently

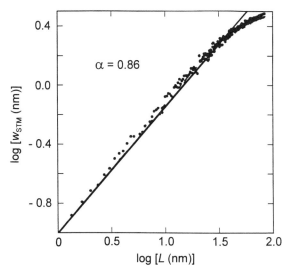

Fig. 6.6 Decimal log–log scaling plot of scanning tunneling microscopy width $w_{STM} \sim L^\alpha$ with surface length L for electrodissolution of (111) silver in 1 M $HClO_4$ in the saturation roughness regime. Current density 30 μA/cm². [Reprinted with permission from M. E. Vela, G. Andreasen, R. G. Salvarezza, A. Hernández-Creus, and A. J. Arvia, *Phys. Rev. B* **53**, 10 217 (1996), Fig. 8, American Physical Society.]

closely follow identical growth patterns. First, recall the study of dendrite formation during electrodeposition of palladium hydride [8], where we discussed the aggregation mechanisms leading to perimeter dimensions of the dendrite forest grown on a strip PdH cathode (Section 3.2.1.2). Now we can appreciate results of the concomitant surface roughness, obtained with the help of Eq. (6.3b) from the imaged width data of the profiles. It turns out that surface roughness exponent α of the PdH dendrites (see Fig. 3.10) amounts to $\alpha = 0.81$–0.83, compared to roughness exponent $\alpha = 0.86$ for Ag surface corrosion discussed above. Second, imaging of electrochemically formed polyaniline films on Pt [9] yielded exponents $\beta = 0.40$, $\alpha = 0.90$, rather close to the above results for Ag corrosion and PdH dendrite formation.

6.1.3 Fractality of Hydrated Cement Paste

Ambient humidity has a significant effect on surface properties of cement paste. For instance, its specific surface increases from about 145 m²/g at 0% relative humidity to about 220 m²/g at 100% relative humidity, with considerable hysteresis between adsorption–desorption steps. Hence, the microstructure of the material, its chemical and physical binding forces, and macroscopic mechanical characteristics are sensitive to the degree of water interaction [10]. Consequently, effects of humidity on readily measurable topologies, such as mass particle and surface element distributions, were explored by fractal analysis on the example of hydrated portland cement paste, using small-angle X-ray scattering (SAXS) over resolution range 20–2000 nm [11].

The SAXS intensities were observed under "slit-smeared" scattering, necessitating augmentation of the scaling exponents of the $I(q) - q$ relations (see Section 3.1.2.2) by term $+1$—equivalent to replacement of intensity $I(q)$ by cumulative intensity $\int I(q)dq$. Figure 6.7 displays a (decimal) log (scattered intensity) − log (wave vector) scaling plot of hardened cement paste [11] for 0% (●) and 100% (○) relative humidity. The numbers in the figure are the numerical least-squares values of the slopes, computed from the data of the various near-linear sections identified by arrows, with standard deviations of about ±0.02.

Table 6.1 lists a summary of the conclusions drawn from the data of Fig. 6.7. Experimental slopes s are collected in column 1, whereas columns 2 and 3 show the ensuing values of the mass fractal dimensions d_M and surface fractal dimension d_S, respectively. These dimensions are calculated [12, 13] via $s = 1 - d_M$ [Eq. (3.8)] or $s = d_S - 5$ [Eq. (3.9)], depending on d_M, d_S falling into its allowed scaling range $1 < d_M < 3$ or $2 < d_S < 3$, respectively (see Section 3.1.3). Table 6.1 further contains the resolution range $R = 2\pi/q$ either side of the observed crossover points (vertical bars in Fig. 6.7).

Separating the outcome according to the length of the linear scale and effects of humidifying, we first look at the smaller yardstick ranges 25–125 and 20–400 nm

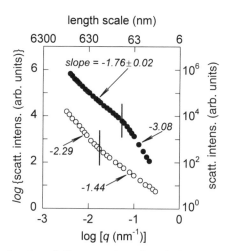

Fig. 6.7 Decimal log–lot plot of slit-smeared small-angle X-ray scattering intensity–wave vector data of oven-dried (●) and hydrated (100% relative humidity, ○) hardened cement paste. Least-squares slopes and their (common) standard deviation are given. [Reprinted with kind permission from D. Winslow, J. M. Bukowski, and J. F. Young, *Cem. Concr. Res.* **25**, 147 (1995), Fig. 3, Copyright 1995, Elsevier Science Ltd., The Boulevard, Langford Lane, Kidlington OX5 1GB, UK.]

TABLE 6.1 Experimental Slopes, their Corresponding Fractal Dimensions, and Relevant Resolution Range $R = 2\pi/q$ for Hardened Cement Paste at 0% and 100% Relative Humidity [11]

Slope[a]	Dimension[b]		Rel. Hum. (%)	R (nm)
	d_M	d_S		
−1.76	2.76	(3.24)	0	125–2000
−3.08	(4.08)	1.92	0	25–125
−2.29	(3.29)	2.71	100	400–2000
−1.44	2.44	(3.56)	100	20–400

[a]From the linear least-squares slopes s of the log–log plots of Fig. 6.7. Here, $d_M = 1 - s$, $d_S = 5 + s$.
[b]Parentheses indicate that the numerical value is forbidden under boundary conditions $1 \le d_M \le 3$, $2 \le d_S \le 3$.

(data rows 2 and 4), noticing that humidifying modifies an essentially Euclidian object ($d_M \approx 3$, $d_S \approx 2$) into a mass fractal ($d_M \approx 2.4$). On the other hand, over the larger scaling ranges up to 2000 nm (data rows 1 and 3), water vapor saturation changes dry cement paste from a mass ($d_M \approx 2.8$) to a surface fractal ($d_S \approx 2.7$). Consequently, humidifying of dry cement paste apparently generates a surface fractal containing smaller domains of mass fractality [11].

If so, the object is a mixed fractal and the simple scattered intensity–wave vector relations established for a purely mass (d_M) or pore (d_P) [Eq. (3.8)] as well as purely surface fractal (d_S) [Eq. (3.9)] no longer hold (see also Table 3.1): According to Eq. (3.14), we must scale the scattered intensity with wave vector q, $I(q) \sim q^s$, under $s = d_S - 2(d_M + d_P) + 6$. Depending on the individual dimensions of the participating fractals, the outcome may indeed simulate the object as surface or mass fractal although the real object cannot be so classified: As we only know resultant exponent s but not the individual d, nothing definite can be said solely on the basis of the data!

The complex results of this study make a particularly strong case for taking assignments of mass and surface fractality, which are solely based on intensity–wave vector scaling, with a pinch of salt. To be convincing, additional methods of fractal analysis must be performed. One or several of the various current adsorbent tiling methods, discussed in Sections 2.2 and 2.3, seems readily applicable to the hardened cement paste. Obviously, application of the theory and procedure of adsorbent preloading by molecular films (Section 2.3.3) is here particularly relevant and promising as humidified cement configurations effectively contain a "film" of water molecules deposited *reversibly* [10, 11] onto the underlying surface elements of their structure.

6.1.4 Representing Shrinkage by Loss of Plasticizer: Julia Sets

Presumably, some readers have used their PC and available software to play with Julia sets, those amazingly complex and beautiful forms that, nevertheless, arise from the simple quadratic iteration process of points z in the complex plane, $z_{n+1} = z_n^2 + c$, under control parameter c. Figure 6.8 shows particular Julia set $J(c = 0.2 + 0.3i)$, generated by 100 000 points of backward orbits [14]. Set $J(c = 0.2 + 0.3i)$ divides all points within the complex plane into two nonoverlapping domains (basins of attraction [15]), namely, the domain of all points that iterate into the attractor (the pentagon-accentuated point $z = 0.079 + 0.357i$), and the domain of all points that iterate to infinity. The jagged boundary of these two domains—in other words, $J(c = 0.2 + 0.3i)$—is chaotic. Points on it are ever undecided which of the two attractors to chose.

Now then, what has such a Julia set to do with fractals in chemistry? Indeed, a Julia set is a convenient tool to *picture* the shrinking perimeter of some material upon its degradation, particularly under a net outgoing flux of plasticizer, water (low humidity), and so on. Changing degrees of shrinkage can be displayed by controlled smoothing of the jaggedness of a Julia set by drawing it increasingly thicker. In addition, its symmetry could be scrambled by introducing random aspects into

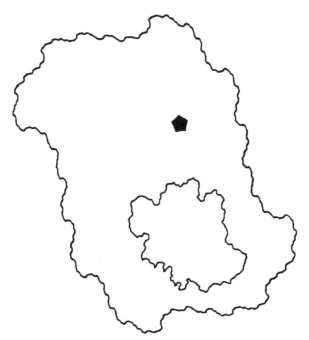

Fig. 6.8 Boundary structure approximating Julia set $z \Rightarrow z^2 + c$, $c = 0.2 + 0.3i$. The real axis (horizontal) reaches from -1 to 1, the imaginary axis (vertical) from $-1.1i$ to $1.1i$. Points on the self-similar boundary are undecided whether to iterate to the point attractor $0.079 + 0.357i$ (center of the pentagon) or to infinity. [Reprinted with permission from R. L. Devaney, in *The Science of Fractal Images*, H. O. Peitgen and D. Saupe, Eds. (Springer, New York, 1988), Chapter 3, Fig. 3.6b, Springer-Verlag.] The inset represents the perimeter of a fresh instant coffee particle (from Fig. 6.1, upper-left inset).

the iteration code. Fractal perimeter dimensions of various stages of generation can be estimated by covering methods (Section 8). To make the point, the silhouette of the instant coffee particle (upper-left inset in Fig. 6.1) has been placed below the attractor in Fig. 6.8.

Of course, the simple quadratic iteration procedure can hardly be taken to represent actual diffusion processes through surfaces and membranes, but neither do other iterative rewriting or production schemes, for instance L-systems generating plant structures [16], account for the actual dynamics of the biological growth processes.

In conclusion, Fig. 6.9 shows a photograph I took of a snow patch on a meadow, under a feeble Michigan February sun. Again, the boundary shape of the patch resembles some connected Julia set, the snow/water crystals at its rim being undecided whether to melt into the darker, hence warmer grass or to condense into the colder, high-reflectivity interior. Figure 6.10 shows the patch hours later. It has broken up into what now resembles a disconnected Julia set [14].

To step outside chemistry here for a brief moment, it would not be difficult to doctor the photo backgrounds such that absolute scales are hidden and then to present what are snow patches as satellite observations of continental cloud covers. So you see, pictures can lie after all!

Fig. 6.9 Photograph of a gradually melting snow patch on a meadow.

Fig. 6.10   Photograph of the snow patch of Fig. 6.9, some hours later.

6.2   A SUMMARY

Observation of surface roughness affords an interesting look into growth kinetic and steady-state textures of interfaces formed during corrosion/dissolution of metal surfaces as well as of its opposite, deposition of film materials on smooth substrate surfaces. Theoretically, by principles of dynamic scaling, and experimentally, using scanning tunneling microscopy, an extension of previously discussed approaches to surface irregularity has been made. We may now analyze nonequilibrium growth kinetics as well as the pattern of the attained steady state, the self-affine interface.

Furthermore, the principles of fractional Brownian motion permit numerical evaluation and modeling of bias in random events, specifically in terms of persistent and antipersistent behavior within the process. Thereby, interface growth following more complex patterns than random diffusion-limited mechanisms can be characterized.

Finally, simple methods of computer generation of certain chaotic boundaries—the Julia sets—may serve as convenient presentations of the structural patterns of shrinking surface perimeters of materials through loss of plasticizer by evaporation, by surface melting, and so on.

NOTES AND REFERENCES

[1] B. J. Barletta, K. M. Knight, and G. V. Barbosa-Cánovas, Compaction characteristics of agglomerated coffee during tapping, *J. Texture Stud.* **24**, 253 (1993); B. J. Barletta and G. V. Barbosa-Cánovas, Fractal analysis to characterize ruggedness changes in tapped agglomerated food powders, *J. Food Sci.* **58**, 1030 (1993). The method used consisted of (a) determining the length L_C of the "coastline" or perimeter (see Section 3.1.2.1) of agglomerated particles, (b) subjecting them to an attrition/compacting process by mechanically tapping of the container, and (c) redetermining L_C at increasing attrition times. Perimeter (or coastline) dimension d_C is obtained via the appropriate log–log scaling plot of perimeter length $L_C \sim \mathcal{N}(\varepsilon)\varepsilon \sim \varepsilon^{1-d}$ [Eq. (1.1)] versus linear yardstick ε, the Feret diameter (Section 3.1.2.1). The graphs indicate good statistics with ε ranging from 0.1 down to 0.01 fractions of Feret's diameter. As expected, a crossover to Euclidian geometry is exhibited; L_C becomes independent of yardstick length once ε falls below 0.01.

[2] D. Saupe, *Algorithms for random fractals,* in *The Science of Fractal Images,* H. O. Peitgen and D. Saupe, Eds. (Springer, New York, 1988), Chapter 2.

[3] M. E. Vela, G. Andreasen, R. G. Salvarezza, A. Hernández-Creus, and A. J. Arvia, In situ sequential STM imaging of structural changes resulting from the electro-dissolution of silver crystal surfaces in aqueous perchloric acid: The roughening kinetics, *Phys. Rev. B* **53**, 10 217 (1996).

[4] M. Aguilar, E. Anguiano, F. Vázquez, and M. Pancorbo, Study of the fractal character of surfaces by scanning tunneling microscopy: Errors and limitations, *J. Microsc.* **167**, 197 (1992).

[5] For computational methods, see [2], Section 2.4.

[6] J. Kertész and T. Vicsek, *Self-Affine Interfaces,* in *Fractals in Science,* A. Bunde and S. Havlin, Eds. (Springer, New York, 1994), p. 96. The general scaling law of the process is postulated as $w(L,t) \sim L^\alpha f(\chi)$, with function argument $\chi = t/L^{\alpha/\beta}$ and α, β independent scaling factors of the self-affine surface. For $\chi \ll 1$ (short times), scaling function approaches power law $f(\chi) \sim \chi^\beta \sim t^\beta/L^\alpha$; hence $w(L,t) \Rightarrow w(t) \sim t^\beta$. For $\chi \gg 1$ (long times), scaling function $f(\chi)$ saturates to a constant; hence $w(L,t) \Rightarrow w(L) \sim L^\alpha$. Scaling exponent α, $0 < \alpha < 1$, is the roughness exponent of the surface (see Glossary 2).

[7] Replace in Eq. (6.1a) t^H of time axis t by x^α of position axis x.

[8] See [37] in Section 3.2.1.2.

[9] M. E. Vela, G. Andreasen, R. C. Salvarezza, and A. J. Arvia, Dynamic scaling analysis of scanning force microscopy images of electrochemically formed polyaniline films in the oxidized form. Scale-dependent roughening kinetics, *J. Chem. Faraday Trans.* **92**, 4093 (1996).

[10] J. J. Völkl, R. E. Beddoe, and M. J. Setzer, The specific surface of hardened cement paste by small-angle X-ray scattering: Effect of moisture content and chlorides, *Cem. Concr. Res.* **17**, 81 (1987).

[11] D. Winslow, J. M. Bukowski, and J. F. Young, The fractal arrangement of hydrated cement paste, *Cem. Concr. Res.* **25**, 147 (1995).

[12] Refer to Eq. (3.8) for a purely mass fractal, to Eq. (3.9) for a purely surface fractal, both their exponents with added term $+1$ from slit smearing, as used specifically in [11].

[13] Numerical dimensions in [11] are given to within 1% of their absolute values but no calculations of error estimates are presented.

[14] The so-called method of encirclement of the Julia set; see H. O. Peitgen, H. Jürgens, and D. Saupe, *Fractals for the Classroom, Part Two. Complex Systems and Mandelbrot Set* (Springer, New York, 1992), p. 378.

[15] See Sections 3.3.1–3.3.4 in [2]. In order not to be misled, it may be useful to add in one's mind the extension "of the attractor" to term "basin of attraction."

[16] P. Prusinkiewicz and J. Hana, *Lindenmayer Systems, Fractals, and Plants* (Springer, New York, 1989).

7

SPECIAL FRACTAL TOPICS IN CHEMISTRY

7.1 FRACTAL DIMENSIONALITY AND UNIVERSAL PROPERTIES WITHIN SYSTEMS

In this chapter, we deal with fractal topics of universality, graph invariants, protein surfaces, macroscopic flow, sorption kinetics, water uptake by paper, turbulent premixed flames, and wear of metal surfaces. These are subjects not included among those discussed in previous chapters but well deserving presentation and discussion. Criteria for their selections is relevance to chemistry and unusual or unexpected fractal characteristics. Of course, there are numerous other special fractal chemistry-related subjects spread throughout the literature. However, many presently lack graphical, computational, or serious experimental support.

We will also discuss instances where the necessary restriction of a fractal scaling application to a finite yardstick range is necessary for a meaningful evaluation of the data. In fact, experimental or inherent difficulties in determining the numerical values of lower and upper cutoffs of the scaling range will be seen to affect the usefulness or predictive capacity of the particular application.

7.1.1 Fractal Dimensions of Coroenes

The term "universality" is often encountered in the literature of fractal phenomena. For instance, we had previously discussed fracton modes and anomalous diffusion, their dimensions $\bar{d}$ offering information on the connectivity—the neighborhood adjacency relations—within a fractal substrate. In turn, the same connectivity may lead to the same spectral dimension $\bar{d}$ [1] and is then said to belong to a "universality class." The advantage of the concept is its generality in describing and cataloguing different materials under a common, more significant heading, thereby avoiding repeat theoretical treatment.

131

Fractals are often generated in a "down" fashion by taking mass away according to a code that results in finer and finer detail on smaller and smaller scales (see Glossary 1, *contraction-shift operation*). Of course, the operation is reversible in the sense that the fractal object can also be built by an "up" process from a single construction element, combining ever more of them according to the generating code: You could not tell from the result which method had been applied, there being no *absolute* scale involved.

A fine chemical example of this is the sequence of multimembered fused benzene rings, the coro[n]enes. Although high *n*-members have not yet been synthesized, the fractal dimension of the self-similar objects can be readily computed and predictions of physicochemical properties of higher stages noted because they approach very closely the (readily computed) properties of the mathematical limit structure, the *attractor* of the underlying iterative code (see Glossary 1). (Note that a prediction of considerable interest here is the number of their per-site Kékulé structures.) To appreciate the structural build-up to higher coro[n]enes, take the benzene molecule as the smallest unit (building block) and combine 6 of them into a hexagonal structure fused along one common C—C bond between neighboring building blocks—which leaves a central hole in the form of the perimeter of the benzene molecule [2, 3]. Now take 6 of these fused 6-benzene structures, consider each as new building block, and arrange them again into hexagonal symmetry by fusing benzene neighbors along a single C—C bond. The perimeter of the central hole now has the form of the rim of coroene. And so on: The series of successive, self-similar stages of formula C_6H_6, $C_{24}H_{12}$, $C_{132}H_{48}$, . . ., approaches a fractal of dimension log 6/log 3 ≈ 1.631 [2–4]. In fact, it works quite well with oyster crackers (and you can eat it too).

Note that this is a deterministic fractal; there are no random contributions in its generating code. Indeed, the coro[n]enes can be much simpler, faster, and near-automatically put on a page by following the prescription of [4]—which constitutes the down generating process. The resulting second stage of the development (the first is benzene) is displayed in Fig. 7.1: The coroene molecule. To see the next stage, coro[2]ene, replace all benzene molecules of coroene by a coroene molecule scaled by linear factor ⅓.

Furthermore, dimension *d* does not depend on any masses or bond length and a common dimension would apply to any class of hexagonal compounds built up of benzene molecules, be it by fusion over an aromatic or by fusion through aliphatic C—C bonds [3]. Similarly, triangular-symmetric molecules approaching the shape of a Sierpinski gasket, the fractal trigonal triphenylenoids, or Clar structures of (not surprisingly) fractal dimension log 3/log 2 ≈ 1.585, are known [5].

As all these structures keep their dimensional relations even if distorted and/or containing random elements, the entire approach serves as zero-order models [2] describing the connectivity of various natural carbonaceous materials—coals, lignites, chars, and soot. Therefore, it seems prudent not to compare experimentally obtained fractal dimensions of such materials exclusively with the predictions of DLA and similar aggregation processes (see Section 3.1.2.2).

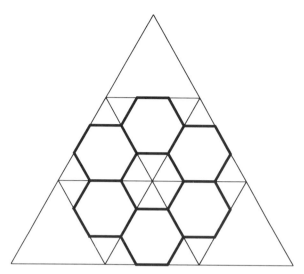

Fig. 7.1 Topological graph of coroene generated as the second stage of a fractal by continually removing self-similar triangular area, scaled by ⅓, from all vertices of equilateral triangles and beginning with an equilateral triangular initiator of unit side length (refer to [4]).

7.1.2 Graph Invariants and Chain Flexibility: *n*-Alkanes and Halogenated *n*-Alkanes

Graph invariants, or topological indexes (TI), are usually in the form of a number, a "distillate" reflecting the size, shape, branching, and number counts of atoms X within some molecular segment or an entire chemical species. Subsequently, quantity TI is related to dependent *macroscopic* molecular quantities such as boiling point, molar volume, and so on, in order to explore and to evaluate functional relationships between dependent and independent quantities.

But what type of relationship? Consider that topology enters naturally into chemistry. Being concerned primarily with how chemical bonds are altered from initial reactant to product, a chemical structural formula of some compound jotted onto the back of an envelope is a topological graph. It is frequently executed within an invariance or universality class because what it intends to show comprises chemical species without regard to their actual bond length and bond angles: It is, again, only the character of *neighborhood relations* that matters mostly. A fractal approach therefore seems rather promising because fractals differ by their connectivity; their actual shape can be pushed and squeezed—as long as nothing tears apart, the fractal dimension remains invariant.

As an introduction, we discuss scaling of the boiling points (K) of the *n*-alkanes from C_1 to C_{40} with the most venerable topological index, carbon number n. Figure 7.2 displays the appropriate (decimal) log–log plot [6]. Note that a common slope

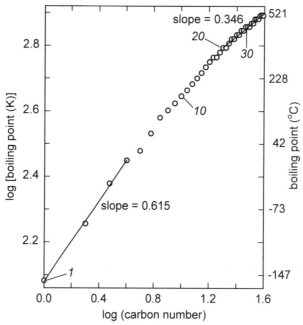

Fig. 7.2 Decimal log–log plot of boiling points versus carbon number of all *n*-alkanes between C_1 and C_{40}. Numbers at data points indicate the carbon number. [Reprinted with permission from D. H. Rouvray and R. B. Pandey, *J. Chem. Phys.* **85**, 2286 (1986), Fig. 2, American Institute of Physics.]

cannot be put through all data points but they appear to imply piecewise slopes among the lowest and the highest carbon numbers, roughly from C_1 to C_{10}, and C_{30} to C_{40}, respectively. In addition, data points begin to crowd more and more at higher carbon numbers n: This already indicates that fractal aspects may be profitably introduced here.

By applying a "property-scale" power law [Eq. (2.17)], we select a useful property to be related to the boiling points and express it as a function of the carbon number n of a series of *n*-alkyl chains. Which physicochemical property *P* of a member of *n*-alkanes can be purposefully related with the temperature of its boiling point? What does a boiling point temperature express in a series of linear chains made up of the same monomer? It obviously relates to the energy to vaporize the liquid—to overcome the attractive forces between neighboring molecules. Certainly, these forces must depend on the *effective* chain length, how molecules are aligned in their liquid-phase conformation. Consequently, we select as property *P* the molecular chain conformation and now have to find a convenient quantity to characterize it.

Recall from Eq. (2.13) that average conformations are characterized by a length measure, say an average radius *R* between chain ends, which scales (a) with carbon number n under exponent *q* (no relation to "wave vector" *q*) as $R \sim n^q$, (b) with mass M under exponent $1/d$ as $R \sim M^{1/d}$ [see Eq. (3.1)], and (c) with the boiling point temperature T under another exponent b, $R \sim T^b$. Therefore, we set

$$P_i \sim R_i \sim n^{q_i} \sim M_i^{1/d_i} \sim T_i^{b_i} \qquad (7.1)$$

where label i designates that all quantities vary over the entire range of carbon number n—as ascertainable from inspecting Fig. 7.2. However, limiting values for low and high n may be assumed for any two n-alkane chain segments of different length, say $i \equiv 1$ for small, index $i \equiv 2$ for large n. Then, (a) we consider a critical chain segment mass M_C such that $M_1 < M_C$ and $M_2 > M_C$, (b) we let M_1, M_2 approach M_C, (c) and get from Eq. (7.1) that

$$T_1^{b_1 d_1} = T_2^{b_2 d_2} \quad M_{1,2} \Rightarrow M_C \qquad (7.2)$$

Furthermore, at limit $M_{1,2} \Rightarrow M_C$, boiling point temperature $T_1 \Rightarrow T_2$ and topological index $(TI)_1 \Rightarrow (TI)_2$. Therefore, $b_1 d_1 = b_2 d_2$ or

$$b_2/b_1 = d_1/d_2 \qquad (7.3)$$

Figure 7.2 demonstrates evaluation of slope q_i/b_i of Eq. (7.1) from least-squares regression analysis over eight successive sets of log (T_i) versus log (n) of data points C_2 to C_{10}, (n small), . . ., C_{32} to C_{40} (n large) [6, 7]. Two resulting slopes, at small n (rigid chain) and large n (flexible chain), respectively, are drawn into the figure. Finally, dividing each slope by initial (short-chain) slope $q_1/b_1 = 0.615$ yields the data entered on the ordinate of Fig. 7.3. Note that numerical value

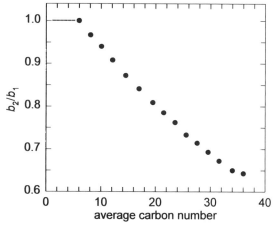

Fig. 7.3 Linear plot of ration b_2/b_1 versus average carbon number of some of the n-alkanes taken from Fig. 7.2. [Reprinted with permission from D. H. Rouvray and R. B. Pandey, *J. Chem. Phys.* **85**, 2286 (1986), Fig. 8, American Institute of Physics.]

$b_2/b_1 = 0.346/0.615 = 0.65$ at n = 36 is considered sufficiently close to theoretical limit $\frac{3}{5} = 0.60$ (see [32] in Section 2.3.5).

In spite of this interesting result, which yields a useful extrapolation procedure for the boiling points of higher carbon-numbered straight-chain aliphatics, it is obvious that carbon number n would not differentiate between straight-chain and branched isomers within an alkane series. This is certainly a disadvantage and a more fundamental topological index is preferably used, the Wiener index W(G) of a topological graph G,

$$W(G) = (\tfrac{1}{2}) \sum_i \sum_j d_{ij} \qquad (7.4)$$

W(G) of G describes adjacency relations between mass points; it is not concerned about actual bond lengths or directions. Demonstrating this for *n*-pentane (Fig. 7.4), we first strip the molecule of its hydrogen atoms and connect carbon atoms by straight lines, nearest neighbors separated by arbitrary unit bond length. Next, we enumerate the adjacency relations among masses (points) of the graph by measuring *all* possible distances d_{ij}, conveniently ordering them by indexes i, j into a row-column array called distance matrix D(G) of graph G. Finally, all distances d_{ij} are summed and the result divided by 2. The scheme in Fig. 7.4, positioned below graph G of *n*-pentane, represents its distance matrix D(G); each of its elements d_{ij}

Fig. 7.4 Topological graph G of *n*-pentane and its distance matrix D(G). Elements d_{ij} (in bold italics) give the adjacency relation of a Manhattan metric, between carbons *i* and *j* labeled by the column and row indexes of D(G). The distance between two nearest neighbors is arbitrarily set to unity.

denotes one adjacency relation between vertices i and j. Computing $W(G)$ from all elements d_{ij} in $D(G)$ by the recipe of Eq. (7.4), we obtain $W(G) = (2 \times 4 + 4 \times 3 + 6 \times 2 + 8 \times 1)/2 = 20$.

Note that Wiener index $W(G)$ gives a number-weighted average of all adjacency relations within a graph; there are more distances at shorter, fewer distances at larger adjacencies—reminding us of fractals. Furthermore, $D(G)$ contains a Manhattan metric (see Glossary 1, *metric*), an outcome appreciated by writing the graph of *iso*-pentane [$W(G) = 17$]. (Carbon number n would not differentiate between the two isomers.) Incidentally, within series of straight-chain homologues it does not matter whether carbon number n or Wiener index $W(G)$ is used; they are related by $W(G) = (n^3 - n)/6$ [8].

Next, we apply these concepts to a scaling of absolute boiling point temperature T with $W(G)$ of straight-chain monobromo-, monochloro-, monofluoro-, and perfluoroalkanes, with the mono-halogen atom at the chain end position [9]. The T–$W(G)$ results are displayed in Fig. 7.5 in terms of fractal chain configuration dimension d_{CH}, instead of ratio b_2/b_1. Recalling that b_1 represents the shortest, stiff alkane chains of $d_{CH} = 1$, we obtain $b_2/b_1 = 1/d_{CH}$ [Eq. (7.3)].

First, note that in Fig. 7.5 we are displaying chain configuration dimension $d_{CH} = 1/b_2$; therefore, the numerical ordinate values of the plot are increasing above unity with increasing $W(G)$ (cf. with Fig. 7.3). Second, recalling Eq. (7.1),

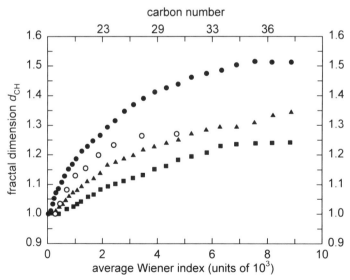

Fig. 7.5 Dependence of chain configuration dimension d_{CH} of monobromo-(■), mono-chloro-(▲), monofluoro-(●), and perfluoro-(○) *n*-alkanes on the Wiener index of the compound. [Reprinted with permission from D. H. Rouvray and H. Kumazaki, *J. Math. Chem.* **7,** 169 (1991), Figs. 5, 6, Baltzer Science Publishers BV, Hooftlaan 51, 1401 EC Bussum, The Netherlands.]

$R \sim M^{1/d}$, we see that dimensions exceeding $d_{CH} > 1$ imply a *shorter* end-to-end Euclidian distance than found for a rigid chain ($d = 1$) of the same carbon number [9], thereby predicting a *more flexible* chain molecule. Third, within the monohalogenated *n*-alkane series, the trend of the data implies that the larger the mass of the halogen atom, the smaller its chain configuration dimension d_{CH} in the limit of large W(G)—hence the more hindered its configurational freedom. For instance, we find $d_{CH} = 1.23$ for bromo and $d_{CH} = 1.52$ for fluoroalkane chains of large Wiener index W(G) or carbon number n.

Although the rigidity of long-chain, unbranched perfluoroalkyls has increased relative to the pure hydrogenated species [cf. d_{CH}(perfluoro) = 1.30 (Fig. 7.5) and d_{CH}(alkyl) = $1/(b_2/b_1)$ = 1.60 (Fig. 7.3)] the perfluorinated molecular chains are still rather pliable: d_{CH}(perfluoro) > 1. In addition, the flexibility of linear long-chain *perfluorinated* alkyl and linear *mono*chloro-alkyl molecules of comparable chain lengths are about equal.

Finally, Fig. 7.6 displays how chain configuration dimensions d_{CH} of perfluorinated *n*-alkanes (from Fig. 7.5) depend on the average molecular volume, up to carbon number n = 20 [9]. As information is not available for n > 16, their molecular volumes were estimated by an extrapolation–regression procedure. The resulting fractal chain configuration dimension d_{CH} = 1.33 for the longest chain (Fig. 7.6) is in agreement with limiting value d_{CH} = 1.30 from boiling point versus Wiener index data (Fig. 7.5).

Summarizing, simple fractal evaluations of basic experimental physical–chemical data within a set of halogen-substituted linear alkyl chain compounds yield useful information on molecular flexibility.

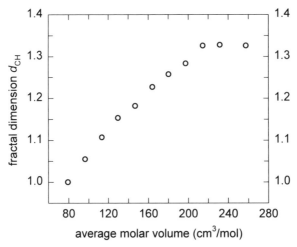

Fig. 7.6 Dependence of chain configuration dimension d_{CH} of perfluorinated *n*-alkanes as function of their average molar volume. [Reproduced with permission from D. H. Rouvray and H. Kumazaki, *J. Math. Chem.* **7**, 169 (1991), Fig. 7, Baltzer Science Publishers BV, Hooftlaan 51, 1401 EC Bussum, The Netherlands.]

7.2 FRACTAL ASPECTS IN BIOCHEMICAL SYSTEMS

We had previously dealt with fractal aspects of systems of interest to biochemists, namely, the aggregate growth of melanin clusters (Section 3.2.2) and hydrogen exchange of lysozyme with surrounding water molecules (Section 4.2.4). We were interested in aspects of diffusion-controlled reaction rates in general and anomalous transport in particular, the nature and properties of the biochemical objects being of lesser concern. We now discuss fractal aspects of biochemical materials with emphasis on their significance to the biochemistry of the observed process.

In Section 7.2.1, previously discussed concepts of fractal surfaces are extended to protein structures, specifically lysozyme. The worthiness—basic and applied—of exploring whether fractality plays a role in protein surface properties is rather obvious, as it is the surface that communicates with the exterior environment via site–receptor interactions, regulates diffusion into and out of the inner protein structure, and so on. Not surprisingly then, studies to look for fractal characteristics of protein surfaces have been undertaken since the early 1980s and interest has been unabated.

Section 7.2.2 deals with effects of aggregation of globin hydrolysate on its self-diffusion mechanism. Fractal aspects are to be expected, particularly if gel formation is involved in the overall process (Section 3.1.1).

7.2.1 Protein Surfaces: Fractality and Receptor Sites

We recall that theories of physisorption consider the adsorbent a nonreactive object and assume minimal adsorbate–adsorbent interaction. Experimentally, analytical tiling methods are preferred that use a homologous series of similar, nonreactive adsorbates (see Fig. 2.2) or apply the inverse tiling approach (Section 2.2) with adsorbate nitrogen. However, surface phenomena of proteins are rather specific in terms of (a) substrate binding sites, which tacitly implies active sites, (b) protein conformations, and (c) any subsequent changes during reactions [10]—surface aspects that incline toward the active site picture of *chemisorption* (Sections 2.4 and 4.3). Therefore, the surface characteristics and dimension of a protein participating in adsorption–desorption phenomena depend, in general, on the *nature* of the adsorbate: Different adsorbates will "see" different protein surface characteristics and properties.

Obviously, the water molecule is an appropriate measure for laying out protein surfaces—we recall its usefulness as interspersed film material in the analysis of surface coverage phenomena of synthetic polymeric materials (Section 2.3.3). We then consider here as surface of a protein the resulting *barrier* between protein and water: Inside this barrier, repulsive interactions exceed 3 kcal/mol; at outside positions, interaction forces are attractive [10, 11].

Figure 7.7 tests these ideas on the lysozyme–water system. The repulsion barrier, shown as thickly rendered boundary, is generated by ab initio molecular orbital calculations using atomic positions from X-ray data, plus a Lennard–Jones potential with added Coulombic term. The contour shown in the figure represents the

20 Å

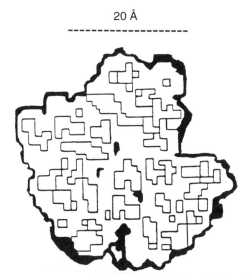

Fig. 7.7   Perimeter of repulsive lysozyme–water interactions, sectioned normal to the length of the protein near its middle. The rectangular shapes denote various amino acids. [Reprinted with kind permission from E. Clementi, G. Ranghino, and R. Scordamaglia, *Chem. Phys. Lett.* **49,** 218 (1977), Fig. 4e, Elsevier Science-NL, Sara Burgerhartstraat 25, 1055 KV Amsterdam, The Netherlands.]

perimeter of a cross section at about middle length of the enzyme [12], the rectangular objects denoting various amino acids [11].

Figure 7.8 gives a (decimal) log–log plot of the number of steps needed to walk this contour at a given step size ε over a range of $1.5 < \varepsilon < 20$ Å. The lower limit accounts for the van der Waals radius of water (1.4 Å), the upper limit reflects the radius of the cross section. The least-squares value of the slope yields contour (perimeter) dimension $d_p = 1.13 \pm 0.02$ [see Eq. (1.1) and Section 3.1.2.1]. The values at the data points in Fig. 7.8 indicate contour length $L(\varepsilon) = \varepsilon^{1-d_p}$, as expected, $L(\varepsilon)$ increases with decreasing ε for $d_p > 1$. Hence, the surface of lysozyme—as seen by water—is a surface fractal of dimension $1 + 1.13 = 2.13$ [10, 13].

Corresponding analysis of eight more, equidistant cross sections of the lysozyme molecule [11] yield fractal dimensions between a low of $d_p = 1.08$ to a high of $d_p = 1.27$, with standard deviations of 0.01–0.03 [10]. Consequently, surface fractality of the lysozyme–water system ranges from $d_s = 1 + d_p = 2.08$ to $d_s = 2.27$.

In a related analysis, a protein–water contact *surface* is generated by surface probing with a spherical particle of the size of the water molecule, $R = R_w$, on the basis of structural information from a protein data bank (Brookhaven). Subsequently, surface analysis evaluates Eq. (2.2) relative to points on the surface, in steps of multiples of log (R_w). The method thus yields *local* as well as *averaged* dimensions of the proteins [14]. Figure 7.9 shows log–log plots of two selected sur-

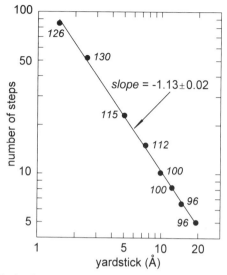

Fig. 7.8 Log–log (decimal) scaling plot of the number of steps of given yardstick size to measure the length of the perimeter of the repulsive protein–water contour of Fig. 7.7, with lengths indicated at the data points. [Reprinted with kind permission from P. Pfeifer, U. Welz, and H. Wippermann, *Chem. Phys. Lett.* **113,** 535 (1985), Fig. 2, Elsevier Science-NL, Sara Burgerhartstraat 25, 1055 KV Amsterdam, The Netherlands.]

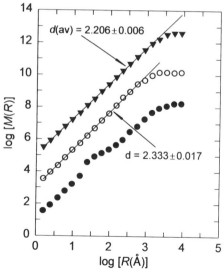

Fig. 7.9 Log–log (decimal) plots of mass–distance scaling of the local surface region near the Leu-6 (○) and the Arg-1 residues (●) of bovine pancreatic trypsin inhibitor and of the protein-averaged measure of lysozyme (▼). [Reproduced with permission from C.-D. Zachmann and J. Brickmann, *J. Chem. Inf. Comput. Sci.* **32,** 120 (1992), Figs. 1, 2, Copyright 1992 American Chemical Society.]

face points of bovine pancreatic trypsin inhibitor, namely, the neighborhoods of the Leu-6 and the Arg-1 residues. Note that the straight slope for Leu-6 implies scaling with surface dimension $d = 2.333 \pm 0.017$. Contrarily, the plot for Arg-1 does not exhibit power law character: Local fractal and Euclidian surface regions are seemingly distributed randomly over the protein surface. Consequently, known active sites on the protein do not consistently exhibit fractal surface characteristics.

Finally, the uppermost data in Fig. 7.9 display the averaged scaling relations over all selected surface points for protein lysozyme, $\langle M(R) \rangle \sim \sum M_i(R_i) \sim \sum_i R_i^{d_i}$. Its slope implies site-average fractal dimension $d = 2.206 \pm 0.006$ [14].

On the basis of these results, the view that surface fractality of proteins is related to the biological functions [10, 15] is shaky; evidence is not propitious. We just saw that fractality of surface regions of bovine pancreatic trypsin inhibitor is distributed in a random fashion over the protein—meaning that regions of specific surface fractality do not coincide with identified, active receptor site locations. Hence, it is hard to maintain that observed protein surface fractality serves a *specific* biological function [14, 16]. On the other hand, the protein–water system is complex and the apparent nonrelevancy of surface fractality of one binding site on the protein out of two is, in fact, not very good statistics.

A somewhat odd aspect of the measured surface fractality is that fractal surface dimensions of the proteins shows relatively low values of $d_S \approx 2.15$. To explain this, the idea was forwarded [10, 17] that a large protein fractal surface dimension favors the rate of capture of water molecules *near* the water–protein interface but disfavors the rate of translational displacements of water molecules *on* the protein surface due to its looping/branching structures (anomalous diffusion; see Section 4.1). A fractal surface dimension slightly exceeding $d_S = 2$ thus constitutes a compromise.

7.2.2 Aggregation and Self-Diffusion of Globin Hydrolysate

We extensively dealt with aggregation phenomena of inorganic and organic materials, the nature of the various substrates being of lesser concern to us; in this section, we will be concerned specifically with aggregation kinetics of globin hydrolysate (from porcine blood cells) as monitored by its self-diffusion coefficient $K(t)$. The $K(t)$ data, recorded at time intervals 0, 1, 2, 3, 4, 5, 6, 8, 10, and 20 (h) by quasielastic light scattering [18, 19], indicate a rapid initial slowing down of the translational mobility of the hydrolysate—particularly accentuated in higher concentrated systems. It is therefore worthwhile to search for possible anomalous diffusion behavior.

We use time-dependent diffusion coefficient $K(t)$ from Eq. (4.3),

$$K(t) \sim t^{-\beta} \tag{7.5}$$

Subsequently, we obtain exponent $-\beta$ from the log–log plot of $K(t)$ versus t. Recording the dependence of β on diffusion time elapsed then yields information on the nature of the diffusion mechanism under the simultaneously progressing clustering reactions of the hydrolysate. In summary, we observe diffusion coefficient $K(t)$ of the diffuser during its own aggregation reaction.

Figure 7.10 displays the $\log[K(t)] - \log(t)$ data of self-diffusion coefficient $K(t)$ of globin hydrolysate at four concentrations ranging from 10 to 25 mg/cm³, as indicated by the italicized numbers attached to the individual slopes. Note that the increasingly negative slopes under increasing globin concentrations imply values of exponent β from 0.08 for the most dilute (10 mg/cm³) down to 0.88 for the most concentrated medium (25 mg/cm³). Although the individual data points scatter considerably, the overall result is reasonable: (a) At low concentrations, hence small aggregation rates, self-diffusion is near-normal [$\beta \approx 0$, $K(t) \Rightarrow K_0$]. (b) Increasing aggregation causes an increasing contribution of anomalous self-diffusion of the globin molecules, as shown by increasing values of exponent β: The diffusing walker is slowed down. (c) At the largest hydrolysate concentration, diffusion is essentially anomalous ($\beta \approx 1$, [19]).

Summarizing, the observed increase of exponent β with increasing initial globin hydrolysate concentrations and with increasing hydrolysate aggregation times implies that self-diffusion of globin hydrolysate becomes successively more anomalous, with β dropping from near regular ($\beta = 0.08$) for the most dilute system to strongly anomalous behavior ($\beta = 0.88$) for the most concentrated solutions. However, we should hesitate to attempt a quantitative assignment of fracton characteristics because the (very likely) presence of *polydispersity* of the globin hydrolysate is not considered [18] (see Section 5.1.2.1 and Glossary 2).

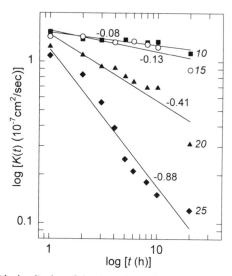

Fig. 7.10 Log–log (decimal) plot of the time-dependent self-diffusion coefficient of globin hydrolysate at room temperature as a function of the aggregation time of the material. The values of the linear least-squares slopes are indicated; italicized numbers give the hydrolysate concentration in (mg/cm³). [Data from X. Q. Liu, M. Yonekura, M. Tsutsumi, and Y. Sano, *J. Agric. Food Chem.* **45**, 1574 (1997).]

7.3 FRACTAL DYNAMICS IN LIQUID–SOLID AND GAS–SOLID FLOWS

In Sections 7.3.1 and 7.3.2 we deal with fractal aspects of the dynamics of water flow in porous soils and of gas–solids two-phase flow in circulating fluidized reactor beds. The first subject is of obvious importance to water–soil research and its many applications to water transport, soil quality, plant harvest yields, and so on. The second subject, circulating fluidized reactor beds, represents a more recent development of fractal analysis to devices for gas–solids operations of catalytic as well as noncatalytic chemical technology.

7.3.1 Horizontal Water Movement in Soils

Application of fractal aspects to experimental data on water movement in soils is of recent origin. Yet in view of the many well-studied fractal phenomena in the related processes of pore adsorption, permeability, and anomalous diffusion in inhomogeneous and fractal substrates, exploring fractal phenomena here is worthwhile and of technological promise.

   In brief, the characteristics of the random horizontal movements of the water content of three natural soils, a Russel silt loam, a Salkum silty clay loam, and a Brazilian clay, were measured by a ^{137}Cs γ-ray attenuation technique [20]. Soil water content w is defined by

$$w = (V - V_i)/(V - V_0) \tag{7.6}$$

where V_i = initial and V_0 = near water-saturated volumetric water content. The random positions $X(t)$ of w are related to fractional Brownian motion [Eq. (6.1a)],

$$X(t) = \lambda(w)t^H \tag{7.7}$$

Just to recall its principle (see Glossary 2): Water transport proceeds by random steps if exponent $H = \frac{1}{2}$. If $H < \frac{1}{2}$, transport is antipersistent—the trend in the random events reverses. If $H > \frac{1}{2}$, transport is persistent—the trend in the random events is maintained. Equation (7.7) is then data fitted by a nonlinear regression analysis.

   Some of the results of water movements in the three soils are shown in Fig. 7.11 by graphing exponent H versus fractional soil water content w. Two limiting situations prevail, depending on high or low water contents. At high fractional water content ($w \approx 1$), water movement in the three soils essentially attains random behavior ($H = \frac{1}{2}$) or slightly persistent characteristics ($H = 0.52$). This outcome is readily understandable: A nearly water-saturated medium offers large regions to liquid transport, enabling it thereby to detour around bottlenecks generated by liquid–solid interfaces. On the other extreme of near-dry conditions, H indicates antipersistent water transport characteristics ($H \approx 0.40 < \frac{1}{2}$), implying that the trend in the random movement $X(t)$ of water volume w at time $t' = t + \Delta t$ is partly

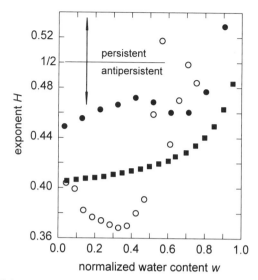

Fig. 7.11 Plot of time exponent H describing horizontal fractional Brownian motion of water content w of a Russel ($\circ$), a Brazilian ($\blacksquare$), and a Salkum ($\bullet$) soil. [Reprinted with permission from I. A. Guerrini and D. Swartzendruber, *Fractals* **2**, 465 (1994), Fig. 1, World Scientific Publishing Co.]

reversed from the trend in $X(t)$ at preceding time $t' = t$. The water movement at low fractional water content retraces its steps (sloshing).

Summarizing, water movement follows a fractional Brownian, antipersistent walk ($H < \frac{1}{2}$) for water-unsaturated soils but attains random walk ($H = \frac{1}{2}$) characteristics near water saturation. Because fractal aspects are scale independent, it is indeed immaterial whether the characteristics of a flux of a macroscopic mass is recorded or the microscopic diffusion of molecules are studied (Chapters 4 and 5): One theory fits all.

7.3.2 Circulating Fluidized Bed Reactors

Next, we explore a fractal approach to the dynamics of a turbulent-flow reactor used in chemical technology. Its working principles are based on two-phase gas–solid flow along the central axis in circulating fluidized beds (CFB) of solid particles of a fluid cracking catalyst (particle diameter 52 μm) and of silica sand (166 μm), respectively. Air at room temperature is used as the fluidizing gas [21].

The CFB systems have been studied in terms of time-averaged values of pressure fluctuations and their standard deviations. In view of chaotic aspects of gas turbulence, the data vary strongly with positions and operating conditions, thereby making it very difficult to obtain a detailed analysis of the working principles of the device. In the present study, a laboratory CFB system is therefore analyzed by applying fractal principles to the recorded time series of the pressure fluctuations of the agitating air stream and of the momentum fluctuations of the colliding solid particles of the reactor beds. Briefly, the method is as follows: (a) The fluctuations of air pressure and particle momentum are recorded separately. Figure 7.12 shows a recording of the pressure fluctuations of the CFB; note the various time scales plotted and the statistical self-similarity of the pattern. (b) The cumulative positions of the air pressure fluctuations are graphed as function of time evolved. (c) The resulting graph is subjected to a box counting method (Chapter 8) to obtain its fractal dimension.

We had previously discussed the evaluation of fluctuating system variables in one form or other; Figure 1.2 demonstrates well the mapping principle of the pressure fluctuations in the CFB reactor by considering 1.2 (*a*) as their descriptor and taking 1.2 (*c*) as the position–time graph of their cumulative positions. More to the point, consider the (self-affine) curve of Fig. 1.4 as a more extensive map of such fluctuations and observe the boxes of size $(b\tau) \times (bc)$ stacked up over a portion of the graph: By performing the box counting procedure, within the resolution range

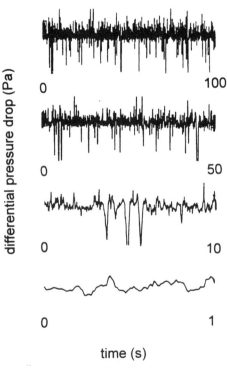

Fig. 7.12   Records of self-similar differential air pressure fluctuations in a circulating fluidized bed reactor. [Reprinted with kind permission from D. Bai, E. Shibuya, N. Nakagawa, and K. Kato, *Powder Technol.* **90**, 205 (1997), Fig. 3, Copyright 1997 Elsevier Science Ltd., The Boulevard, Langford Lane, Kidlington OX5 1GB, UK.]

inherent in the actual position–time map, a box counting dimension d (of the map) of the pressure fluctuations within the CFB reactor is thereby obtained.

Pertinent results are collected in Table 7.1. Column 1 lists the fractal (box counting, local) dimension d, column 2 displays exponent $H = 2 - d$ (see Glossary 2, *fractional Brownian motion*), and columns 3 and 4 list operative parameter values. For instance, note that box counting dimension d decreases with decreasing gas-flow rates under otherwise identical operating conditions, implying more uncorrelated system states ($H \Rightarrow \frac{1}{2}$). Second, if the circulation rate of the solids is raised with other conditions unchanged, d also decreases and H approaches unbiased random behavior ($H = \frac{1}{2}$).

Undoubtedly, a fractal approach to analyze the working characteristics of gas flow and confined particle motions during operation of such complex-chaotic reaction devices is quite useful. The system also furnishes a fine example of time-series evaluations, applications of fractional Brownian theory to quantify biased random walks, and of the relation of its time exponent H to the fractal dimension of the self-affine position–time graph of the observed random variable, the reactor's gas flow.

7.4 HYDROCARBON SORPTION IN CROSS-LINKED POLYMERIC SYSTEMS

Diffusion and sorption of aliphatic and aromatic hydrocarbons in cross-linked polymers and elastomers is of considerable interest as the presence of such plasticizers within the lattice causes swelling and influences shear relaxation. On a more general level and for many other polymer networks, this is of relevance to gel permeation chromatography, ion exchange, and controlled release of drugs. Many interesting papers are extant on this subject, displaying numerous experimental results from evaluations of the data by the classical laws of diffusion (see Section 4.1) in terms of the amount of penetrant sorbed, $M(t)$, versus the square root $t^{1/2}$ of the duration t of the sorption process [22]. The plots do not always exhibit convincingly

TABLE 7.1 Fractal Dimension d and Exponent H of Fractional Brownian Motion Under Some Operating Conditions of a Circulating Fluidized Bed Reactor of Height 0.4 m [21]

Fractal Dimension d	Exponent H	Solid Circulation Rate $(kg\ m^{-2}s^{-1})$	Gas Flow (m/s)
1.71	0.29	8	2.5
1.62	0.38	30	2.5
1.60	0.40	42	2.5
1.63	0.37	8	1.5
1.55	0.45	12	1.5
1.50	0.50	18	1.5

straight lines and even if they do it is not immediately apparent whether their slopes s would indeed correspond to $s = \frac{1}{2}$. The obvious was to replot some of the data in unbiased log–log fashion. In fact, fractal aspects of some scope came to light, implying complex hydrocarbon sorption kinetics in vulcanized natural rubber.

7.4.1   Hydrocarbon Sorption by Vulcanized Natural Rubber

7.4.1.1   Aliphatics n-C$_6$ to n-C$_9$

We discuss the sorption data [23] in terms of cumulant penetrant uptake $M(t) \equiv M_t$ versus sorption time t by plotting $\log (M_t)$ versus $\log(t)$ in order to get an unprejudiced idea on the reality of classical penetrant diffusion kinetics in the vulcanized rubber samples. We therefore assign the resulting slope s to exponent H of fractional Brownian motion, $M_t \sim B_H(t) \sim t^H$ [see Glossary 2 and Eq (6.1a)].

The series of sorption experiments are parametrized by temperature, type and degree of vulcanization (hence nature and density of rubber cross-links), and chemical nature of the hydrocarbon penetrant or permeate [23]; it is helpful to label the makeup of the (four) selected data bases in a penetrant–adsorbent system *matrix*:

Penetrant	Adsorbent	°C	Label	
n-C$_6$H$_{14}$	Conventional and peroxide-vulcanized to 34-dNm torque	28	(S1)	
n-C$_9$H$_{20}$	Conventional and peroxide-vulcanized to 34-dNm torque	28	(S2)	(7.8)
n-C$_9$H$_{20}$	Conventional and peroxide-vulcanized, optimum cured	28	(S3)	
n-C$_9$H$_{20}$	Peroxide-vulcanized to 34 dNm	60	(S4)	

The term conventional refers to vulcanization by sulfur (generation of long polysulfide bridges connecting the chains of rubber molecules) and dNm signifies the units of the testing moment of force in deci-Newton meter $= 10^{-1}$ m^2 kg s^{-2}.

Figure 7.13 displays the $\log [B_H(t)] - \log(t)$ plots of system label S1 of Eq. (7.8); penetrant uptake $B_H(t)$ is given in moles per 100 g elastomer sample. Equilibrium values B_∞ of penetrant uptake are marked at the right-side ordinate axis; horizontal bars specify their half-values, $\log[\frac{1}{2} B_H(t \Rightarrow \infty) \equiv \frac{1}{2}B_\infty]$, the apparent upper limit of linear scaling behavior prior to saturation [24]. Note that $H > \frac{1}{2}$ within narrow experimental errors (± 0.02 or better, standard deviation): Sorption of hexane is *persistent* under $H \approx 0.63$ (see Glossary 2, *fractional Brownian motion*). Note also that equilibrium n-hexane uptake is larger by conventionally vulcanized than by dicumyl peroxide-vulcanized rubber, the former process causing large internal spaces, the latter process inducing direct C—C linkage between rubber chains [23].

Figure 7.14 displays the results of the data evaluations for heavier aliphatic n-nonane [system S2 of Eq. (7.8)]. Again, penetrant diffusion of the higher molecular weight molecule proceeds under essentially the same persistent characteristics ($H \approx 0.64$) as found for smaller n-heptane penetrant (Fig. 7.13). Understandably, equilibrium uptake B_∞ of n-nonane is less than of n-hexane, but ration $R_\infty =$

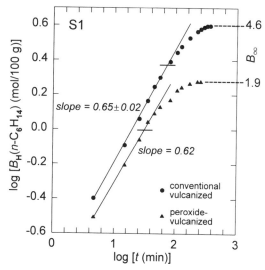

Fig. 7.13 Log–log (decimal) plot of the position–time graph of fractional Brownian motion of *n*-hexane during sorption at 28 °C by natural rubbers vulcanized to a torque of 34 dNm. System label of Eq. (7.8) is indicated in the left upper quadrant. [Reproduced by permission from G. Unnikrishnan and S. Thomas, *J. Polym. Sci.* B **35**, 725 (1997), Fig. 2, Copyright 1997 John Wiley & Sons, Inc.]

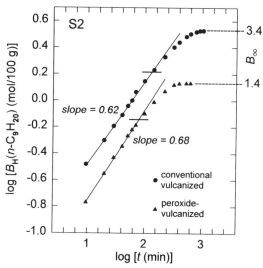

Fig. 7.14 Log–log (decimal) plot of the position–time graph of fractional Brownian motion of *n*-nonane during sorption at 28 °C by natural rubbers vulcanized to a torque of 34 dNm. System label of Eq. (7.8) is indicated in the left upper quadrant. [Reproduced by permission from G. Unnikrishnan and S. Thomas, *J. Polym. Sci.* B **35**, 725 (1997), Fig. 3, Copyright 1997 John Wiley & Sons, Inc.]

B_∞(sulfur)$/B_\infty$(peroxide) = 2.42 for sulfur- and peroxide-vulcanized rubbers is the same as for penetrant n-hexane, R_∞ = 2.43 (Fig. 7.13).

Figure 7.15, system label S3 of Eq. (7.8), indicates there is no discernible difference in the values of H if a rubber specimen is cured beyond the common torque of 34 dNm, all other conditions remaining the same (see System label S2).

Figure 7.16 (system label S4) displays the effects of increased temperature during the sorption process of n-nonane in peroxide-vulcanized rubber. Comparing the data with the appropriate 28 °C data (▲) for system label S2 in Fig. 7.14, we notice that the sorption process at the higher temperature has a larger degree of persistent character: $H = 0.77$ at 60 °C versus $H = 0.68$ at 28 °C, a significant increase outside experimental errors. In addition, B_∞(60 °C) ≈ B_∞(28 °C), attesting that H indeed reflects the *kinetics*.

What can we conclude from these results? Essentially, they imply that the sorption dynamics of the alkanes n-heptane and n-nonane in vulcanized rubbers proceed by persistent fractional Brownian walk. Hence, the a priori assumption of classical diffusion behavior [22, 23] is not verified by the data. Conceivably, there are rubber lattice–penetrant interactions that drive the penetrant mobilities into persistent-biased random characteristics (accelerated diffusion).

To extend such notion further, assume that a penetrant diffusing into cross-linked rubber sees a Euclidian space. Hence, each diffusing coordinate describes the map

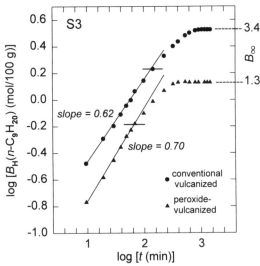

Fig. 7.15 Log–log (decimal) plot of the position–time graph of fractional Brownian motion of n-nonane during sorption at 28 °C by optimally vulcanized natural rubbers. System label of Eq. (7.8) is indicated in the left upper quadrant. [Reproduced by permission from G. Unnikrishnan and S. Thomas, *J. Polym. Sci. B* **35**, 725 (1997), Fig. 5, Copyright 1997 John Wiley & Sons, Inc.]

of an independent fractional Brownian motion process which, over some time interval Δt, traces a length ΔL of

$$\Delta L \sim (\Delta t)^H \tag{7.9}$$

Neglecting overlap, the self-similar *trail* (or *path*) of the diffusing penetrant generates a mass M scaling with elapsed time Δt as $M \sim \Delta t$ [Eq. (3.18)]. Therefore, $M \sim (\Delta L)^{1/H}$ [Eq. (7.9)] $\sim (\Delta L)^d$ [Eq. (2.2)]. We conclude that the dimension of the trail, d_T, is related to exponent H by $d_T = 1/H = 1/0.62 = 1.6$ [25–27].

Noting numerical value $1/H = 2$ for random walk in spaces $E \geq 2$—the trail of a Brownian random walker eventually fills its space completely—the sorption results $1/H = d_T = 1.6$ predict that fractional Brownian motion of the n-alkyl penetrants in vulcanized rubber fills available space *less* completely than by random penetration. Evidently, this is a corollary of the persistent, "forward-driving" diffusional solvent motion at $H > \frac{1}{2}$ within the rubber lattice.

7.4.1.2 Benzene and Higher Aromatics

It is, of course, most interesting to extend the sorption studies to aromatics: Indeed, identical sorption experiments and fractal evaluation procedures were performed at 28 °C with the aromatics benzene, toluene, p-xylene, and mesitylene [28]. Figure 7.17 displays $\log[B_H(t)]$–$\log(t)$ scaling of the sorption of these penetrants by

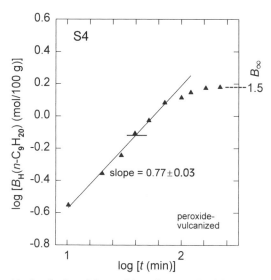

Fig. 7.16 Log–log (decimal) plot of the position–time graph of fractional Brownian motion of n-nonane during sorption at 60 °C by natural rubber vulcanized to a torque of 34 dNm. System label of Eq. (7.8) is indicated in the left upper quadrant. [Reproduced by permission from G. Unnikrishnan and S. Thomas, *J. Polym. Sci.* B **35**, 725 (1997), Fig. 8, Copyright 1997 John Wiley & Sons, Inc.]

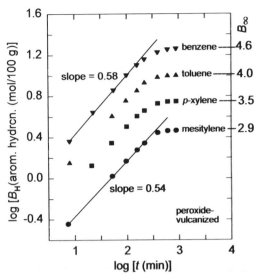

Fig. 7.17 Log–log (decimal) plot of the position–time graph of fractional Brownian motion during sorption of four aromatics at 28 °C by natural rubber vulcanized to a torque of 34 dNm. Individual plots are arbitrarily up-shifted for better viewing. [Reproduced with kind permission from G. Unnikrishnan and S. Thomas, *Polymer* **35**, 5504 (1994), Fig. 8, Copyright 1994 Elsevier Science Ltd., The Boulevard, Langford Lane, Kidlington OX5 1GB, UK.]

dicumyl peroxide-vulcanized natural rubber; the individual plots for p-xylene, toluene, and benzene are shifted upward for better viewing. Note, first, that exponent $H \approx 0.56$ is common to the four aromatics within experimental uncertainty and, second, that it is closer to random value $H = \frac{1}{2}$ than established for the n-aliphatics sorbed by identically vulcanized rubber (see Figs. 7.13 and 7.14). Consequently, we expect that the dimension of the trail of the diffuser is closer to $d = 2$ of unbiased random walk, as discussed at the end of Section 7.4.1.1. Indeed, we find $d_T = 1/H = 1.8$.

Figure 7.17 also shows that the equilibrium sorption concentrations B_∞ of the aromatic molecules are comparable to those of the alkyl compounds.

Finally, we apply fractal analysis to the data of sorption experiments of toluene in conventionally vulcanized natural rubber (polysulfide cross-links) containing fillers of silica and carbon black, respectively, recalling that both fillers are usually in the form of DLA or percolation-type aggregates (see Sections 3.1.2.1 and 3.1.2.2). Figure 7.18 displays the resulting scaling plot of penetrant uptake $B_H(t)$ versus elapsed sorption time t. Exponent H and fractal dimension of the diffuser's trail, $d_T = 1/H$, are essentially independent of the nature of the filler within the resolution ranges of these experiments. Comparison with Fig. 7.17 indicates that the addition of these fillers does not affect exponent H within one standard deviation.

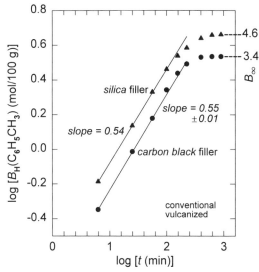

Fig. 7.18 Log–log (decimal) plot of the position–time graph of fractional Brownian motion of toluene during sorption at 28 °C by natural, filler-added rubber vulcanized to a torque of 34 dNm. [Reproduced with kind permission from G. Unnikrishnan and S. Thomas, *Polymer* **35**, 5504 (1994), Fig. 7, Copyright 1994 Elsevier Science Ltd., The Boulevard, Langford Lane, Kidlington OX5 1GB, UK.]

7.4.2 Sorption Kinetics of Aromatics in Urethane-Modified Bismaleimide Elastomers

Here we deal with sorption of benzene, toluene, and *p*-xylene by a network of cross-linked urethane-modified bismaleimide elastomers [29]. We thereby explore how chemical elastomer structures of a widely different nature from those of vulcanized rubbers affect the kinetics of the sorption process of the same penetrant molecules. Preparatively, first a polyurethane prepolymer is synthesized by reacting 4,4′-diphenyl methane diisocyanate with poly(oxytetramethylene) glycol (PTMO) of molecular weight 1000 and 2000, respectively. Thereafter, the product is condensed with maleic anhydride under a $1:2$ molar ratio. One maleic acid group is thus attached at either end of the prepolymer via $>O + OCN{\sim}R{\sim}NCO + O< \Rightarrow CO_2 + >N{-}R{-}N<$, (UBMI), with R designating the PTMO chains. Subsequently, the UBMI are cross-linked over the maleimide double bonds, the process initiated by benzoyl peroxide. In the following, we denote these elastomers by UBMI–PTMO1000 and UBMI–PTMO2000, respectively.

The sorption experiments are performed according to the procedures mentioned in Section 7.4.1.1; the appropriate scaling plots log $[B_H(t)]$–log(t) are displayed in Fig. 7.19 for both elastomers and common penetrants benzene, toluene, and *p*-xylene. There is no significant variation in exponent H with penetrant species as

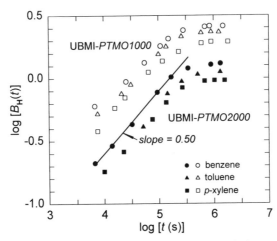

Fig. 7.19 Log–log (decimal) plot of fractional Brownian walk during sorption of benzene, toluene, and *p*-xylene at 25 °C by two urethane-modified bismaleimide elastomers. The straight line is the least-squares analysis of the benzene data. [Reproduced by permission from D. L. Liao, Y. C. Chern, J. L. Han, and K. H. Hsieh, *J. Polym. Sci.* B **36**, 1747 (1997), Fig. 5, Copyright 1997 John Wiley & Sons, Inc.]

well as with CH_2-chain lengths of the elastomer (UBMI–PTMO1000, open symbols and UBMI–PTMO2000, closed symbols): $H = \frac{1}{2}$ implies Brownian motion, in contrast to the sorption kinetics of penetrant toluene in vulcanized natural rubber (Section 7.4.1.2), where H exhibits significant persistent character ($H = 0.58$, Fig. 7.17).

Summarizing, the absorption of hydrocarbon penetrants by the elastomers described in the subsections of Section 7.4 is within the range of persistent ($H > \frac{1}{2}$) to Brownian random behavior ($H = \frac{1}{2}$): Antipersistent characteristics ($H < \frac{1}{2}$) are not discernible under given conditions. Specifically, exponent H does not vary for a given elastomer structure with the size of the related penetrant molecule. Furthermore, addition of filler materials has no influence on the values of H.

Obviously, this type of data evaluation contains considerably greater predictive power than a priori assumption of $t^{1/2}$ sorption kinetics which, effectively, merely serve to 'confirm' classical behavior. There is a definite dependence of exponent H of sorption kinetics on the cross-linking characteristics for the natural rubber adsorbent: Polysulfide-linked chains induce a lesser degree of persistent diffusion than chain molecules cross-linked by shorter C—C bonds. Furthermore, under otherwise identical conditions, increasing the temperature of the sorption process leads to an increased degree of persistent character of the fractional Brownian motion of the penetrant species.

What to make, in detail, of the numerical values of H and $1/H$ in terms of anomalous but *accelerated* diffusion behavior? Foremost, a great deal more experimentation is required under closer-spaced $B_H(t) - t$ sampling intervals toward an improved accuracy of the fractal evaluations but it seems worth the effort. In addition, the principles of topological indexes discussed in Section 7.1.1 can be combined with sorption kinetics to explore commonalities of prevalent penetrant–elastomer interaction potentials which ought to give some information on the influence of molecular flexibility within a series of chemically identical but structurally different penetrant molecules.

7.5   WETTING AND POROSITY: THE WATER–PAPER TOWEL EXPERIMENT

Uptake of liquids by paper and by paper products is of widespread occurrence in chemical technology, such as analytical methods of chromatography, in drying processes, and in color deposition and application methods. In addition, the dynamics of such wetting processes are seen to follow principles and theories of self-affine surface roughening relevant to other, entirely different chemical topics: For example, metal surface corrosion by acid (discussed in Section 6.1.2) and photoablation of polymer surfaces (to be dealt with in Section 7.8). This relationship furnishes further demonstration how a fractal approach, once quantified, can be of wide applicability to a variety of different chemical–physical topics; fractal analysis emphasizes the interrelation and commonality of, on first sight, entirely different processes.

The laboratory study of the dynamics of paper-wetting uses a simple setup [30]. A paper towel is placed flatly onto a plate glass at an angle of 60° to the horizontal. The plate contains a reservoir at its bottom, filled with red-dyed water; the edge of the paper towel dips by about 1 cm into the colored liquid. Subsequently, the up-moving water front is photographed at 10 different time intervals between 0 and 3602 s, each color print is scanned in black and white for improved contrast, and subsequently digitized over a total length of 240 mm into 10 points/mm, or $x \leq L = 2400$. Average paper–water interface heights $\langle h(L,t) \rangle = L^{-1} \sum h(x,t)$, generated by sorption of water by the pores and locally different densities of the paper structure, are read from the digitized photographs.

By recalling our discussion on dissolution kinetics in Section 6.1.2, we can expect that the root-mean-square width $w(L,t) = (\langle h^2 \rangle - \langle h \rangle^2)^{1/2}$ of the water–paper interface is self-affine, scaling with time as $w(t) \sim t^\beta$ and, after maximum penetration is attained, scaling with length L as $w(L) \sim L^\alpha$ [31]. However, numerical evaluation is somewhat complex as $w(L,t)$ of the water–paper interface indicates irregular behavior for $t < 30$ s from some initial, transient flow through the pore system of the paper. At much longer times ($t > 2500$ s) the dynamics slow down due to some smoothing of the interface roughness. Hence, the peak values of $w(L,t)$ at $t = 2500$-s process time, which are parametrized by 17 values of L between 200 and 2200, are selected to represent $w(L)$ at saturation interface growth conditions.

Figure 7.20 depicts the (decimal) log–log plot of $w(L)$ versus basis length L, thereby yielding surface roughness exponent α at saturation water uptake. Note that for short-length regions, $L < 120$ mm, $\alpha = 0.67 \pm 0.04$; for long-length regions, $\alpha = 0.19$. The crossover into a range of large L scaling with a much smaller exponent

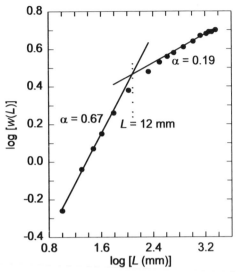

Fig. 7.20 Log–log (decimal) plot of interface width $w(L)$ versus interface basis length L for water sorption by a paper towel. The crossover region is indicated by the dotted line. [Reproduced with permission from T. H. Kwon, A. E. Hopkins, and S. E. O'Donnell, *Phys. Rev. E* **54**, 685 (1996), Fig. 4, American Physical Society.]

$\alpha = 0.19$ implies complex dynamics of water sorption by the paper towel, and a much rougher interface. Because surface roughness is probed under condition $L \Rightarrow 0$ [31], we only consider the short-L data region ($L \leq 12$ mm) as a valid indication of the surface roughness of the paper–water interface.

Growth exponent β in $w(t) \sim t^{\beta}$ is obtained by scaling $w(L,t)/L^{\alpha}$ versus $t/L^{\alpha/\beta}$ [31] and fitting β to yield the best straight slope in the corresponding log–log plot: $\beta = 0.24 \pm 0.02$ [30]. This value compares with growth exponent $\beta = 0.36$ for solvated acid-induced roughening of a (111) Ag surface (Section 6.1.2) and with $\beta = 0.37$ for solution–solid interface growth during Ag electrodeposition (Section 3.2.1.1).

It is useful to emphasize here that dynamic scaling with time t (growth exponent β) and static scaling with basis length L (roughness exponent α) of the self-affine water–paper interface represent local events that are restricted to short distance/time intervals relative to size of the object and duration of the sorption process.

7.6 FRACTAL PARAMETERS OF PREMIXED TURBULENT LEAN FLAMES

We discuss now a relatively recent and active pursuit of fractal applications and analysis to a particular field of chemistry, namely, combustion in turbulent premixed flames (wrinkled flame regime).

The *modus operandi* of the process is described in terms of asymptotically thin moving laminar flamelets imbedded in turbulent flow. As their instantaneous behavior is the same as that of laminar flames, the product of turbulent flame surface area A_T and laminar burning velocity S_L yields the turbulent burning velocity S_T, $S_T = S_L A_T / A_L$—a number of obviously great interest; not surprisingly, the turbulent flame surface turns out to be a self-affine fractal [32, 33]. Hence, we shall now consider how A_T is measured. The flame parameters we need for discussion [34] are listed in Eq. (7.10):

A_L Surface area of a laminar flame (length^2)

A_T Surface area of a turbulent flame (length^2)

S_L Laminar burning velocity (length s^{-1})

S_T Turbulent burning velocity (length s^{-1}), the mass consumption
      rate per unit area and reduced to density of the unburned
      mixture   (7.10)

u' Turbulence intensity, the root-mean-square fluctuation of the
      flame velocity (length s^{-1})

R_E Reynolds number, $R_E = u'L/\nu$, L = turbulent integral length,
      ν = kinematic viscosity

ϕ Air/fuel equivalence ratio

Figure 7.21 shows a minimal schematic [35] of a test section for studying turbulent flame fronts, indicating the turbulence-inducing grid, premixed air–fuel mixture

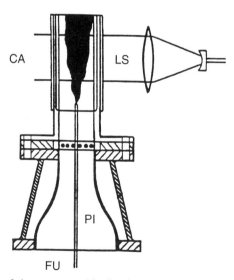

Fig. 7.21 Schematic of the test assembly for the camera recording of a turbulent flame front. LS = laser sheet, CA = camera, ····· = turbulence-inducing grid, PI = pilot burner, FU = air–fuel mixture inlet. [Reproduced with permission from A. K. Das and R. L. Evans, *Expts. in Fluids* **22,** 312 (1997), Fig. 4, Springer-Verlag.]

intake, pilot burner, argon ion laser sheet, and recording camera (shutter speed 1/8000 s). The camera image comprises an area of $x = 45$, $y = 60$ mm of the laser sheet of central width of 0.250 mm. Photographs are scanned at 60 pixel/mm, equivalent to 12 pixel/mm on the scale of the flame (5:1 optical reduction). Resolution is therefore 0.083 mm per pixel [36].

Figure 7.22(a) shows the photograph of a premixed, lean ($\phi < 1$) methane–air turbulent flame, as looked at from its side [36]. The lightened regions represent the unburnt sections of the flame, rendered visible by scattered light from TiO_2 particles (decomposed from initially injected $TiCl_4$). The flame itself is delineated as the dark portion due to decreased laser light-scattering (high-temperature induced drop in TiO_2 particle density).

The interface between the two regions is the flame front. Note its irregularity; an image-processed picture of such flame front, or perimeter of a flame surface, is displayed by Fig. 7.22(b). Figure 7.23 demonstrates rather well the effects of increasing turbulence—quantified by increasing turbulence intensity u'/S_L [see Eq. (7.10)] from left to right—on the examples of three flame fronts, with noticeable fragmentation of the flame front at highest u'/S_L [32].

The digitized images of such flame fronts are then subjected to fractal analysis,

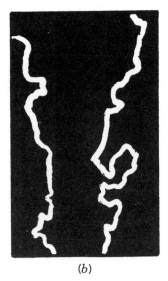

(*a*)   (*b*)

Fig. 7.22 (*a*) Film-recorded image of a turbulent flame viewed from its side. The bright region is the unburnt fuel–air mixture, delineated by laser light scattered from TiO_2 particles. The interface between dark and bright regions is the self-affine flame front. (*b*) Instantaneous position of a self-affine flame front after image processing; the flame front has been artificially thickened. [Reproduced with permission from A. K. Das and R. L. Evans, *Expts. in Fluids* **22**, 312 (1997), Figs. 7, 8b, Springer-Verlag.]

usually by scaling their length $L(\varepsilon) \sim \mathcal{N}(\varepsilon)\varepsilon \sim \varepsilon^{1-d_P}$ with yardstick size ε [see Eq. (1.1)] between a lower and upper limit, $\varepsilon_i \leq \varepsilon \leq \varepsilon_0$, called inner and outer cutoff. Slope s of the log $[L(\varepsilon)]$–log(ε) plots, $s = 1 - d_P$, with d_P the perimeter dimension of the flame front, then leads to surface fractal dimension $d_S = 1 + d_P$ [37], $d_S = 2 - s$.

Figure 7.24 shows a typical scaling plot of ensemble-averaged length $\langle L(\varepsilon) \rangle$ of a premixed methane–air flame of $\phi = 0.6$, $u' = 0.17$ m/s, $u'/S_L = 1.4$, and $R_E = 38$ [36], observed under the setup shown in Fig. 7.21. Therefore, from slope $s = -0.098$ one gets flame surface dimension $d_S = 2.1$.

Figure 7.25 depicts scaling of length $L(\varepsilon)$ of the image of a propane–air flame front of a Bunsen burner of $\phi \approx 0.9$ and $u' = 0.47$ m/s [38, 39]. In this system, we obtain flame surface dimension $d_S = 2.27$ from slope $s = -0.27$. Note, first, that the inner cutoff does not saturate at $\varepsilon_i \ll \varepsilon$, a general phenomenon not yet satisfactorily understood [40]. Second, we see that fractal surface dimensions d_S of turbulent premixed flame fronts vary from about 2.10–2.27—a range variability that is not surprising considering different flame conditions, burner types, nature of fuel, fuel–air equivalence ratio, and so on.

Finally, Fig. 7.26 shows a comparison between the methane–air [36] and the propane–air flame [38] of Figs. 7.24 and 7.25 in terms of their respective surface dimension d_S at common degrees of turbulence—expressed by ratio u'/S_L [see Eq. (7.10)]. Noticeably, the fractal surface dimension of a flame increases with increasing turbulent intensity, rising from its lower, laminar flame limit of $d_S \approx 2$.

As mentioned briefly at the beginning of this section, the considerable technological significance of area measurements of turbulent (A_T) and laminar (A_L) flames is that they offer a method for estimating the burning velocity of a turbu-

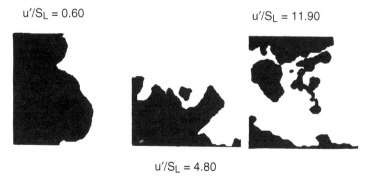

Fig. 7.23 Images of typical flame fronts at indicated relative turbulence intensity u'/S_L. [Reprinted with permission from G. L. North and D. A. Santavicca, *Comb. Sci. Technol.* **72**, 215 (1990), Fig. 6, Gordon and Breach Publishers, World Trade Center, 1000 Lausanne 30, Switzerland.]

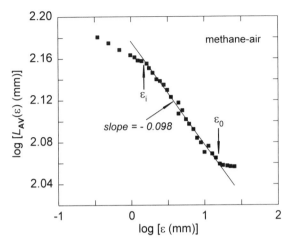

Fig. 7.24 Log–log (decimal) scaling plot of the ensemble-averaged perimeter length of the turbulent front of a lean premixed methane–air flame. Inner and outer cutoff points are indicated. Slope s corresponds to $s = 1 - d$, with d = fractal dimension of the flame front. [Reproduced with permission from A. K. Das and R. L. Evans, *Expts. in Fluids* **22**, 312 (1997), Fig. 9, Springer-Verlag.]

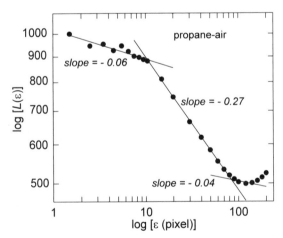

Fig. 7.25 Log–log (decimal) scaling plot of pixel perimeter length versus pixel yardstick size of the front of a turbulent flame of a Bunsen burner. Slope s corresponds to $s = 1 - d$, with d = fractal dimension of the flame front. [Reprinted with permission from A. Yoshida, M. Kasahara, H. Tsuji, and T. Yanagisawa, *Combust. Sci. Technol.* **103**, 207 (1994), Fig. 2, Gordon and Breach Publishers, World Trade Center, 1000 Lausanne 30, Switzerland.]

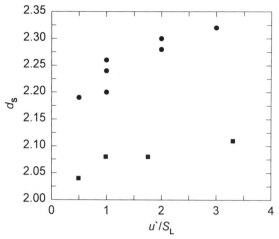

Fig. 7.26 Linear plot of the fractal surface dimension of the flame fronts of premixed lean turbulent methane–air (●) and propane–air (■) flames as function of their relative turbulent intensity u'/S_L. See also Figs. 7.24 and 7.25, respectively. [Data from A. K. Das and R. L. Evans, Expts. in Fluids **22**, 312 (1997), and A. Yoshida, M. Kasahara, H. Tsuji, and T. Yanagisawa, *Combust. Sci. Technol.* **103**, 207 (1994).]

lent flame front S_T, relative to that of the laminar flame front, S_L; $S_T/S_L = A_T/A_L$ [34, 38]. Expressing laminar and turbulent flame front areas according to Eq. (2.1),

$$A_L(\varepsilon_i) = \varepsilon_i^2 \varepsilon_i^{-d_s} = \varepsilon_i^{2-d_s}$$

$$A_T(\varepsilon_0) = \varepsilon_0^2 \varepsilon_0^{-d_s} = \varepsilon_0^{2-d_s} \qquad (7.12)$$

we find that ratio S_T/S_L depends on three experimentally accessible parameters: The turbulent flame surface dimension d_S and the inner and outer cutoffs ε_i and ε_0,

$$S_T/S_L = (\varepsilon_i/\varepsilon_0)^{(2-d_S)} \qquad (7.13)$$

We have then a clear example that the restriction of a fractal analysis to a *finite linear scaling* range is here a *fundamental* necessity in order to use the very fractal approach [41]. Without precise information on actual values of inner and outer cutoff, it is impossible to compute ratio S_T/S_L and thereby obtain the burning velocity of a turbulent flame front. Usually, ratio $\varepsilon_i/\varepsilon_0$ is of the order of 0.1, as can be read off Fig. 7.24 or 7.25. By using, for instance, $d_S = 2.20$, Eq. (7.13) predicts $S_T/S_L \approx 1.6$.

Summarizing, we see that application of fractal aspects to premixed turbulent lean flames has made it possible—although ambiguities remain—to estimate turbulent flame burning velocities on the basis of a few experimental parameters pertaining to the surface area of the turbulent flame front. This is a considerable accomplishment in view of the fact that application of Euclidian geometry failed [38].

7.7 BIFRACTALITY: WEAR BETWEEN PROCESSED METAL SURFACES

Here, we consider fractal surface aspects of frictional wear between metal surfaces [42]. There is little need to stress the importance of wear through point-to-point contact of nominally flat surfaces in all aspects of engineering involving moving machinery (including the flesh-and-blood type). The wear model is particularly interesting as it introduces bifractality, meaning that two fractal dimensions are generated from different mechanistic aspects of the surface-forming processes. Thus, more than one discernible fractal dimension, each valid in its own scaling range, is required for a proper description of the overall phenomenon.

Briefly, the surfaces of alloy 83Sn/11Sb/6Cu (Babbitt alloy) are ground or lapped by an abrasive and the result is analyzed by profilometry [42, 43]. The radius of the stylus tip amounts to 2 μ; computer-sampling spacings of $\vartheta \approx 1$ μ of the digitized data are chosen [44] under common sampling length $L = 2.343$ mm. Figure 7.27 displays two views of the profiles of the digitized surface irregularities

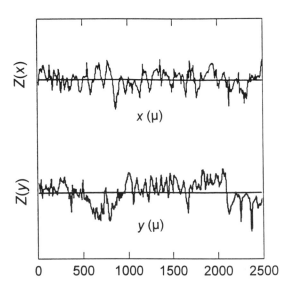

Fig. 7.27 Experimental profiles $Z(x)$, $Z(y)$ of a lapped Babbit alloy surface taken in random horizontal direction x and y (normal to x) over the surface. Maximal size of the fluctuations are of the order of ± 1 μ about zero level. The computer-sampling interval is $\vartheta = 0.78$ μ over total length $L = 2343$ μ. [Reprinted with kind permission from L. He and J. Zhu, *Wear* **208**, 17 (1997), Fig. 2a, Copyright 1997 Elsevier Science Ltd., The Boulevard, Langford Lane, Kidlington OX5 1GB, UK.]

$Z(x)$—the interface between the object and space above—in direction x and direction y normal to length coordinate x. The point-by-point surface irregularity of heights $Z(x)$ about mean $Z(x) = 0$ and at horizontal coordinates x are then digitized and evaluated as follows. For N such points we select *lag distance* $\Delta = n\vartheta$ ($n = 1$, 2, . . .) starting at points n within the total profile length L and under computer sampling spacing ϑ of the digitized $Z(x)$ data. Subsequently, semivariance (structure function) $G(\Delta)$ of the $Z(x)$ is computed,

$$G(\Delta) = (½)\langle[Z(x + \Delta) - Z(x)]^2\rangle \tag{7.14}$$

The data accumulation relation is conveniently modified to [42]

$$G(\Delta) = [1/(N - n)]\sum_{j=0}^{N-n}(Z_{j+n} - Z_j)^2 \tag{7.15}$$

expressing sampling of the momentary profile heights $Z(x_j) \equiv Z_j (j = 0, 1, 2, . . . , N)$ at equidistant lag distance Δ over length L. Finally, plotting log $[G(\Delta)]$ versus $\log(\Delta)$ yields slope $s = 2H$, and hence fractal profile dimension $d_p = 2 - H$ (see Glossary 2, *fractional Brownian motion*) via

$$G(\Delta) \sim \Delta^{2H} \sim \Delta^{4-2d_p} \tag{7.16}$$

Figure 7.28 displays the results. Note the appearance of *two* scaling regions at profile dimension $d_p = 2 - s/2 = 2 - 1.59/2 = 1.23$ (region 2) and $d_p = 1.62$

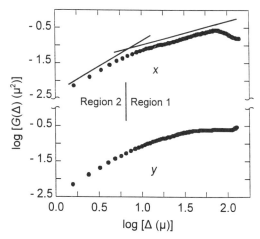

Fig. 7.28 Log–log (decimal) scaling plot of structure function $G(\Delta)$ versus lag distance Δ of the surface profiles of Fig. 7.27. Slopes s, indicated by straight lines, relate to surface profile dimension d by $d = 2 - s/2$. Region 1: $s = 0.73$, $d = 1.62$. Region 2: $s = 1.59$, $d = 1.23$. [Reprinted with kind permission from L. He and J. Zhu, *Wear* **208**, 17 (1997), Fig. 2b, Copyright 1997 Elsevier Science Ltd., The Boulevard, Langford Lane, Kidlington OX5 1GB, UK.]

(region 1, $s = 0.73$). In particular, surface dimension $d_S = 1 + 1.62 = 2.62$ of region 1, over horizontal scaling range $6 < L_1 < 30$ μ, is said to arise from plastic surface deformations exerted by the abrasive material. On the other hand, surface dimension $d_S = 2.23$ of region 2, $0 < L_2 < 6$ μ, is said to represent statistically self-similar microfractures arising from microcutting, microploughing, and microrolling by the hardened abrasive lapping particles beyond the plasticity limit of the metal alloy surface [42].

In summary, the object and its surface roughening processes discussed here are perhaps more of a chemical engineering than of chemical interest but they furnish a good example of bifractal behavior with a problem-related rationalization.

7.8   PHOTOABLATION OF POLYMER SURFACES

Another method of surface roughening, particularly apt for polymer molecules possessing smooth surfaces, uses a laser beam of wavelength comparable to chemical bond energies. Thereby, direct photoexcitation under low thermal effects—which may cause surface relaxation/ordering—is attained. Technologically such methods, called photoablation, are quite promising, particularly for etching the extremely nonreactive and smooth surfaces of polyfluoro polymers.

We dispense with discussions of the experimental methods of data collection and the theoretical basis of their evaluation and presentation, as they have been dealt with just above, and thus concentrate on the results of an application of this approach to roughen the originally smooth surfaces of polyimide and polytetrafluoroethylene [45]. We only have to keep in mind that *surface* images rather than *profile* images (see Section 7.7) are scanned and evaluated. Consequently, scaling relation Eq. (7.16) is modified to

$$G(\Delta) \sim \Delta^{2(3 - d_S)} \tag{7.17}$$

because the random walk now covers a two-dimensional space under $E = 2$: $H = 3 - d_S$ [46].

The resulting fractal surface dimensions are $d_S = 2.59$ for polytetrafluoroethylene and $d_S = 2.79$ for polyimide over scaling range 1–60 pixel. The smaller fractal surface dimension d_S of the tetrafluoroethylene polymer is ascribed to a relatively lower stability to laser-induced thermal surface relaxation processes [45].

7.9   SURFACE ROUGHNESS AND SURFACE FRACTALITY

We have now discussed several different processes of surface roughening. It may be therefore helpful to put the fractal aspects of surface probing of metal lapping and polymer photoablation into proper perspective with electrolytic roughening of a (111) Ag surface or the water–paper interface of the paper towel wetting experiment, treated in Sections 6.1.2 and 7.5, respectively. Note that (a) here we obtained *surface fractal* or surface *perimeter* dimension d_S or d_P, (b) there we obtained a *surfac roughness* exponent α. Admitting that the analytical methods as well as the theory are related, how is a surface roughness exponent $0 < \alpha < 1$ to be juxtaposed to a surface perimeter dimension $1 < d_P < 2$? Furthermore, can we set $H = \alpha$?

The answers are found by looking carefully at the nature of the quantities measured. Recall, first, that in Section 6.1.2 we evaluated the root-mean-square width of Ag surface height increments. The latter were taken as the averaged-reduced roughness coordinates in a *vertical* direction to the rough surface (refer to the zoomed-in sketch in Fig. 6.3). Hence, we measure a (frequently self-affine) quantity characterizing the roughness of the surface and not a statistically self-similar fractal surface or fractal profile dimension. Indeed, theory of the appropriate scaling function [Eq. (6.3b)] requires us to go to limit $L \Rightarrow 0$; roughness exponent α is a *local* quantity. The same arguments hold for the water–paper towel experiment discussed in Section 7.5; in fact, they were already briefly outlined there. Recall that the closer roughness exponent $\alpha \Rightarrow 1$, the *smoother* is the surface.

On the other hand, in Sections 7.7 and 7.8, we consider the structure factor [Eqs. (7.16) and (7.17)] at many lag distances Δ, taken over the observed length L of the alloy surface—sort of a moving average. Consequently, we measure a (frequently statistically self-similar) *less* local fractal profile or fractal surface dimension.

These are the principles. Regarding whether roughness exponent α equals exponent H of fractional Brownian motion, it would also depend on the method of data evaluation; usually $H \approx \alpha$ in many experimental situations [47, 48]. In fact, we have used this relation in Section 6.1.2 for evaluating roughness results of a corroded surface of silver and of aggregated dendrites of palladium hydride, respectively. Indeed, that exponent H and roughness exponent α are closely related can be readily ascertained by considering that strongly antipersistent behavior, $H \Rightarrow 0$, yields a very jagged profile because each random trend essentially undoes the action of its predecessor. In complete parallel, we also find $\alpha \Rightarrow 0$. In the other extreme, a strongly persistent fractional Brownian motion process, $H \Rightarrow 1$, leads to a rather smooth, near-Euclidian profile of dimension $d \approx 1$. Similarly, roughness exponent $\alpha \approx 1$ characterizes a surface near devoid of rough features.

However, keep in mind that roughness exponent α refers to limit $L \Rightarrow 0$, where L is the extent or length of the probed surface, and therefore represents a local quantity (see above). Hence, its valid scaling range should, a priori, not involve larger values of inner and outer cutoffs than that of exponent H of fractional Brownian motion.

NOTES AND REFERENCES

[1] See [11] in Section 4.2. The universality is only strictly valid in the limit of infinite structures.

[2] D. J. Klein, T. P. Živković, and A. T. Balaban, The fractal family of coro[n]enes, *Match Comm. Math. Chem.* **29,** 107 (1993). Counting begins with benzene as $n = 0$.

[3] D. J. Klein, M. J. Cravey, and G. E. Hite, Fractal benzenoids, *Polycycl. Arom. Compds.* **2,** 163 (1991).

[4] Some years ago, Professor Freeman drew this fractal to be distracted on a boring flight. Upon his return, we figured out a simple construction code. (a) Draw a large, precise equilateral triangle. (b) Mark each side length into equal thirds. (c) Number the marks, say in the mathematical positive sense (opposite the clock), as 2, 3, 5, 6, 8, 9, calling the vertices of the triangle 1, 4, 7. (d) Draw a line between marks 2 and 9, between 3 and 5, and between 6 and 8. (e) Discard the triangles at the three corners: A hexagon remains, the graph of the benzene molecule.

Now divide this hexagon into its six equilateral triangles and do the same with *each* triangular piece as you did with the larger initiator. In other words, *each* side of the six triangles is divided into three identical segments according to (b) and (c), the markings are connected by straight lines as in (d), and all three corner triangles of each triangle are discarded as in (e): The graph of the coroene molecule has been generated; its central hole equals the perimeter of benzene. It is displayed in Fig. 7.1.

Fractal dimension. First stage ($n = 1$): 3 triangles are discarded, 6 remain of length scale $\frac{1}{3}$ each. Second stage ($n = 2$): 6×3 triangles are lost, 6×6 *remain,* of length scale $\frac{1}{9}$ each. Third stage ($n = 3$): $6^2 \times 3$ triangles discarded, $6^2 \times 6$ *remain,* of length scale $\frac{1}{3^3}$ each. Hence [Eq. (1.1)], self-similarity dimension $d = -\log(6^n)/\log[(\frac{1}{3})^n] = \log 6/\log 3 = 1.63\cdots$. Larger than topological dimension $d = 1$, because we continually add smaller and smaller lines, smaller then topological dimension $d = 2$, because we continually discard smaller and smaller area. Linear range of the stages: From the length of the benzene molecule up to the expertise of the synthetic chemist. An interesting corollary of $d = 1.63$ is that density of the neat planar compound, $\rho \sim R^d/R^2 \sim R^{d-2} \sim R^{-0.37}$, drops to rather low values at higher stages [3].

[5] D. Plavšić and N. Trinajstić, Clar structures in fractal benzeneoids, *Croat. Chem. Acta* **65**, 279 (1992). It is not necessary to show a figure of the Sierpinski gasket as it is in nearly all books on fractal theory and graphics.

[6] D. H. Rouvray and R. B. Pandey, The fractal nature, graph invariants, and physiochemical properties of normal alkanes, *J. Chem. Phys.* **85**, 2286 (1986).

[7] Slopes are determined by least-squares regression analysis over a sliding set of no less than eight adjacent carbon numbers, with the initial slope value (n small) taken as denominator for all others. Ratios b_2/b_1 are thus decreasing from unity with increasing TI.

[8] D. H. Rouvray, (a) The modeling of chemical phenomena using topological indices, *J. Comput. Chem.* **8**, 470 (1987); (b) *The role of the topological distance matrix in chemistry,* in *Mathematics and Computational Concepts in Chemistry,* N. Trinajstić, Ed. (Horwood, Chichester, UK, 1986), p. 295; A. T. Balaban, *Numerical modeling of chemical structures: Local graph invariants and topological indices,* in *Group Theory and Topology in Chemistry,* R. B. King and D. H. Rouvray, Eds. (Elsevier, Amsterdam, 1987), p. 159.

[9] D. H. Rouvray and H. Kumazaki, Prediction of molecular flexibility in halogenated alkanes via fractal dimensionality, *J. Math. Chem.* **7**, 169 (1991).

[10] P. Pfeifer, U. Welz, and H. Wippermann, Fractal surface dimension of proteins: Lysozyme, *Chem. Phys. Lett.* **113**, 535 (1985).

[11] E. Clementi, G. Ranghino, and R. Scordamaglia, Intermolecular potentials: Interaction of water with lysozyme, *Chem. Phys. Lett.* **49**, 218 (1977).

[12] A. P. Allen, J. T. Colvin, D. G. Stinson, C. P. Flynn, and H. J. Stapleton, Protein conformation from electron spin relaxation data, *Biophys. J.* **38**, 299 (1982). In this prefracton work, the dimensional quantity in the exponent of the density of vibrational states was assigned to a fractal rather than to spectral dimension $\bar{d}$ (see Glossary 2).

[13] For the relation to obtain surface dimension d_s from its perimeter dimension d_p, $d_s = 1 + d_p$, see Section 1.2 and Fig. 1.5.

[14] C.-D. Zachmann and J. Brickmann, Hausdorff dimension as a quantification of local roughness of protein surfaces, *J. Chem. Inf. Comput. Sci.* **32**, 120 (1992). The yardstick was $1.2 < R < 20$ Å. The measured dimensions are referred to as "Hausdorff" dimensions (see Glossary 1) but amount, effectively, just to simple coverings by spherical test volumes of known dimension. To prove that an exponent p of a covering is a Hausdorff dimension, a priori upper and lower limits of the Hausdorff p-dimensional measure need to be taken; see Gerald A. Edgar, *Measure, Topology, and Fractal Geometry* (Springer, New York, 1990), p. 152.

[15] J. Åqvist and O. Tapia, Surface fractality as a guide for studying protein–protein interactions, *J. Mol. Graph.* **5**, 30 (1987).

[16] Refer also to [23] in Section 4.2.4.

[17] P. Pfeifer, D. Avnir, and D. Farin, Scaling behavior of surface irregularity in the molecular domain: From adsorption studies to fractal catalysts, *J. Stat. Phys.* **36**, 699 (1985).

[18] J. E. Martin and F. Leyvraz, Quasielastic-scattering linewidths and relaxation times for surface and mass fractals, *Phys. Rev.* A **34**, 2346 (1986).

[19] X. Q. Liu, M. Yonekura, M. Tsutsumi, and Y. Sano, Quasi-elastic light scattering study on the globin hydrolysate gel formation process, *J. Agric. Food Chem.* **45**, 1574 (1997).

[20] I. A. Guerrini and D. Swartzendruber, Fractal characteristics of the horizontal movement of water in soils, *Fractals* **2**, 465 (1994).

[21] D. Bai, E. Shibuya, N. Nakagawa, and K. Kato, Fractal characteristics of gas–solids flow in a circulating fluidized bed, *Powder Technol.* **90**, 205 (1997). The log–log scaling plots of the box counting procedure are not shown here; they are linear over a range of $0.02 < t < 82$ s, with slight deviations from linearity to shortest times due to noise.

[22] P. Neogi, *Transport Phenomena in Polymer Membranes,* in *Diffusion in Polymers,* P. Neogi, Ed. (Dekker, New York, 1996), p. 173.

[23] G. Unnikrishnan and S. Thomas, Sorption and diffusion of aliphatic hydrocarbons into crosslinked natural rubber, *J. Polym. Sci.* B **35**, 725 (1997). The analytical procedures of sorption of straight-chain aliphatics by vulcanized natural rubber are singularly simple, using nothing more sophisticated than a balance for weighing samples in the range of grams and, perhaps, a torque meter to verify the degree of vulcanization of the rubber samples. In brief, disk-shaped rubber specimen are soaked in the respective solvent, removed at predetermined times, quickly cleaned of adhered solvent (by toweling), then weighed and subsequently put back into the solvent.

[24] Beyond $B_{1/2}$ the analysis of exponent H from the experimental scaling slopes is no longer valid since solvent saturation is approached; see [22].

[25] R. F. Voss, *Fractals in nature: From characterization to simulation,* in *The Science of Fractal Images,* H. O. Peitgen and D. Saupe, Eds. (Springer, New York, 1988), Chapter 1, p. 64.

[26] See Chapter IX, Section 27 of [2], cited in Section 1.1.

[27] Keep in mind the difference between the *trail* and the *map* of a (fractional) random walk; see Section 1.2 for an employ of the map of a random walk on $E = 1$.

[28] G. Unnikrishnan and S. Thomas, Diffusion and transport of aromatic hydrocarbons through natural rubber, *Polymer* **35**, 5504 (1994).

[29] D. L. Liao, Y. C. Chern, J. L. Han, and K. H. Hsieh, Swelling equilibrium and sorption kinetics of urethane-modified bismaleimides elastomer, *J. Polym. Sci.* B **36,** 1747 (1997).

[30] T. H. Kwon, A. E. Hopkins, and S. E. O'Donnell, Dynamic scaling behavior of a growing self-affine fractal interface in a paper-towel-wetting experiment, *Phys. Rev. E* **54,** 685 (1996). It is noteworthy that the roughness exponent of the interface of a (dry) paper tear amounts to $\alpha = 0.66$; see T.-H. Kwon, The scaling behavior of a self-affine fractal interface in a paper tearing experiment, *Fractals* **5,** 121 (1997).

[31] See [6] of Section 6.1.2.

[32] G. L. North and D. A. Santavicca, The fractal nature of premixed turbulent flames, *Comb. Sci. Technol.* **72,** 215 (1990).

[33] See p. 100 in [2] of Section 1.1.

[34] I. Glassman, *Combustion* (Academic, Orlando, FL, 1996).

[35] There is no place here to display the extensive hardware equipment needed for combustion studies. For the relative simpler setup used in laminar flame research, see E. W. Kaiser, W. G. Rothschild, and G. A. Lavoie, Effect of fuel–air equivalence ratio and temperature on the structure of laminar propane–air flames, *Comb. Res. Technol.* **33,** 123 (1983).

[36] A. K. Das and R. L Evans, An experimental study to determine fractal parameters for lean premixed flames, *Expts. in Fluids* **22,** 312 (1997).

[37] See (a) pp. 213–214 in [2] of Section 1.1, (b) Section 1.2 and Fig. 1.5 of this text.

[38] A. Yoshida, M. Kasahara, H. Tsuji, and T. Yanagisawa, Fractal geometry application in estimation of turbulent burning velocity of wrinkled laminar flame, *Comb. Sci. Technol.* **103,** 207 (1994).

[39] Flame surface dimension d_S is rendered in the combustion literature by symbol D_3.

[40] Inner cutoff ε_i is related to the interaction of the turbulent eddies with the flame front; turbulent eddies smaller than ε_i are bypassed by the flame without interaction; see Ö. L. Gülder and G. J. Smallwood, Inner cutoff scale of flame surface wrinkling in turbulent premixed flames, *Comb. Flame* **103,** 107 (1995).

[41] O. Malcai, D. A. Lidar, O. Biham, and D. Avnir, Scaling range and cutoffs in empirical fractals, *Phys. Rev. E* **56,** 2817 (1997). The possibility is raised that scaling ranges over less than two decades are system inherent. The paper chronologically lists all reports on fractals in the *Physical Review* journals from Jan. 1990–Dec. 1996.

[42] L. He and J. Zhu, The fractal character of processed metal surfaces, *Wear* **208,** 17 (1997). The study mentions the Weierstrass–Mandelbrot fractal function [M. V. Berry and Z. V. Lewis, On the Weierstrass–Mandelbrot fractal function, *Proc. Royal Soc. London Ser.* A **370,** 459 (1980)] but makes no apparent use of its special features except for its simple structure function (its trend under slowest varying contributions), which is proportional to $G(\Delta) \sim \Delta^{2H}$ [see Eq. (6.1b)].

[43] The "lap", a rotating disk for polishing a gem or metal. (The Oxford Encyclopedic English Dictionary, Oxford University Press, New York, 1995.)

[44] A continuous signal $f(v)$, with a Fourier transform $FT[f(v)] = F(u)$ band-limited at $-u_b$ and $u_b [F(u) = 0$ for $|u| \geq u_b]$, is reconstituted by pulses $f(n\vartheta)$ at a sequence of equidistant points $\Delta = n\vartheta$ if $\vartheta \leq 1/(2u_b)$; see A. Papoulis, *Systems and Transforms with Applications in Optics* (Krieger, Melbourne, FL, 1981), p. 119.

[45] Cs. Beleznai, R. Vajtai, and L. Nánai, UV-Induced Fractal Surfaces, *Fractals* **5,** 275 (1997).

[46] The relation between exponent H and fractal dimension d of the walker's map is $H = E + 1 - d$, where E is the Euclidian dimension of the space walked. Here, $E = 2$, hence $H = 3 - d$; see [25], Chapter 1, p. 45.

[47] L. A. Bursil, F. XuDong, and P. Julin, Fractal analysis of crystalline surfaces at atomic resolution, *Philos. Mag.* A **64,** 443 (1991). See also J. M. William and T. P. Beebe, Jr., Analysis of fractal surfaces using scanning probe microscopy and multiple-image variography. I. Some general consideration; II. Results on fractal and non-fractal surfaces, observation of fractal crossovers, and comparison with other fractal analysis techniques, *J. Phys. Chem.* **97,** 6249 (I), 6255 (II) (1993).

[48] T. Jøssang and J. Feder, The fractal characterization of rough surfaces, *Phys. Scr.* *T***44,** 9 (1992).

8

FRACTALITY AND
ITS MEASUREMENTS

8.1 THE BOX COUNTING METHOD

In this book, we discussed a variety of experimental methods of fractal analysis of chemical systems, such as adsorption/tiling measurements, reaction rate determinations, scattering techniques, dispersive spectroscopies, electron microscopies, scanning tunneling microscopy, nuclear magnetic spin–echo measurements, and image analysis.

But how do we obtain the fractal dimension of some random object on a graph, of an image, or a molecular drawing? We had discussed the perimeter scaling method, measuring the length of a fractal "coastline" by stepping around it under varying step or "divider" size, of such diverse fractal objects as silica clusters on one hand and instant coffee particles on the other. Another, more general measuring technique is the so-called box counting method. It is frequently used because it can be readily automated or even performed manually—albeit considerably more tediously. Furthermore, its principle is transparent and easily understood.

The box counting or capacity measure [1] is based on scaling principles of Eq. (1.1): (a) A cubic lattice for three-dimensional or a grid in the case of in-plane projections, of lattice constant (box size) ε, is laid over the object to be measured. (b) The number of boxes, $N_B(\varepsilon)$, which cover *any* part of the object (the "occupied" or "intersected" boxes), are counted and each data couple $N_B(\varepsilon)$, ε is tabulated. (c) Procedure (b) is repeated with a set of successively smaller ε. (d) Log $[N_B(\varepsilon)]$ is plotted versus log $(1/\varepsilon)$ or log (ε) and slope s of the resulting straight line—if such indeed exists—taken as the box counting dimension $d = s$ or $d = -s$, respectively, of the object.

Note that the box counting method somewhat resembles the Hausdorff measure: In both methods an object is covered by generalized balls and their diameter is successively decreased. In fact, the Hausdorff measure does not use equal-sized but arbitrarily sized balls of diameter diam and volume $(diam)^p$, where p is an a priori unknown exponent, $0 \leq p < \infty$. In addition, its limiting procedures are more stringent (see *covering*, Glossary 1). Indeed, the box counting dimension need not coincide with the Hausdorff dimension, nor with the numerical values from other definitions of dimension [2] such as the perimeter scaling method, nor is it necessary that the object is self-similar—although this is not a distinguishing property; it holds equally well for Hausdorff and perimeter scaling methods.

8.1.1 Dimension of a Branching Molecule: Dendrimer DAB(CN)$_{64}$

We apply the box counting method to dendrimer molecule DAB(CN)$_{64}$, a three-dimensional, near-spherical poly(propylene imine) structure of molecular weight 6910 (theoretical) and of 10960-Å^3 volume [3], sketched in Fig. 8.1. Dendrimer DAB(CN)$_{64}$ is a tree molecule, constructed from n-propylene segments that branch into one additional segment at each three-valent N atom. The (partial) dendrimer in Fig. 8.1 represents the fourth branching state of the development [4], with three additional identical trees anchored at the three remaining bonds of the two nitrogen atoms of the central diaminobutane group, making up the complete DAB(CN)$_{64}$ structure. In other words, we only require the graph of the neighborhood relations of the dendrimer molecule (see Sections 7.1.1 and 7.1.2). The exploration of its fractality, possibly arising from the particular branching pattern, then obviously

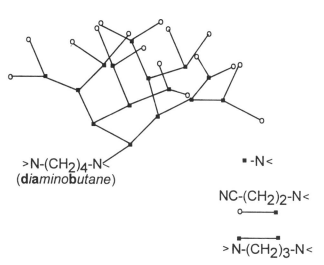

Fig. 8.1 Planar projection of poly(propylene imine) dendrimer molecule DAB(CN)$_{64}$. Only one of the four identical-branched structures, anchored at the nitrogen atoms of the central diaminobutane group, is shown. Bond distances and angles are schematized, methylene group chains $(CH_2)_3$ and $(CH_2)_2$ rendered as line segments, nitrogen branching points by ■, and the end-standing CN group depicted by ○.

does not require covering of the three-dimensional object; it suffices to project its structure onto the plane and use, conveniently, rectangular- or disk-shaped cover elements. Furthermore, we do not have to be concerned with actual bond angles or distances as long as we keep the lines representing $(CH_2)_3$ groups connecting the N-atom branching points of equal length. It may also be prudent to draw the structure avoiding a large degree of overlap.

Figure 8.2 demonstrates the box counting procedure on the projection of one of the four identical tree structures of $DAB(CN)_{64}$ under grid size $\varepsilon = 10$, of which $N_B(\varepsilon = 10) = 88$ grid boxes are occupied or intersected by line elements of the dendrimer. Figure 8.3 displays the effect of decreasing grid size ε from 10 to 2.5. Now there are $N_B(\varepsilon = 2.5) = 366–369$ occupied boxes. Note the coincidence of a segment of the object with a grid line, pointed out by the vertical arrow. (Reorientation of the projection would eliminate this particular ambiguity.)

Figure 8.4 shows the resulting $\log[N_B(\varepsilon)] - \log(\varepsilon)$ plot with box size $\varepsilon = 1.67$, 2, 2.5, 2.86, 3.33, 4, 5, 10, and 20, of which size 1.67, 2, 2.5, 2.86, and 4 were counted several times. Obviously, the data points do *not* lie on a straight line but on a nonlinear (parabolic) regression curve, as indicated. On the other hand, a strictly linear least-squares regression analysis yields slope $s = -0.974 \pm 0.02$ for the range of smaller box sizes, $1.67 \le \varepsilon \le 4$, and $s = 1.20$ for the largest box sizes, $\varepsilon = 5$, 10, and 20. Certainly, larger values of ε violate condition $\varepsilon \Rightarrow 0$; ensuing dimension $d = 1.20$ is therefore less reliable.

What to make of this result? First, this one-shot procedure needs improvement. In order to get good statistics, we ought to repeat the box counting with the den-

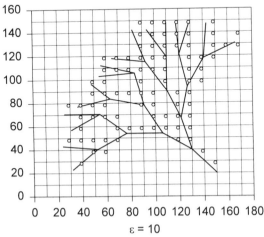

$\varepsilon = 10$

Fig. 8.2 Planar projection of dendrimer molecule $DAB(CN)_{64}$ within a near-quadratic grid of arbitrary box size $\varepsilon = 10$. Only one of the four identical branched structures of the molecule is shown and their common diaminobutane anchor is omitted (see Fig. 8.1). Bond angles and distances are not meant to be realistic. Occupied (intersected) boxes are marked by a circle in their upper right corner.

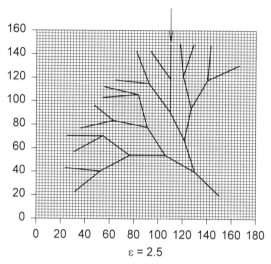

Fig. 8.3 Planar projection of dendrimer molecule $DAB(CN)_{64}$ within a near-quadratic grid sixteen times denser than that of Fig. 8.2. The vertical arrow points to a molecular segment falling onto a grid line.

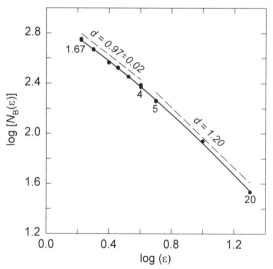

Fig. 8.4 Decimal log–log plot of the number of occupied boxes versus nine values of linear box size, $1.67 \leq \varepsilon \leq 20$, applied to the projection of dendrimer molecule $DAB(CN)_{64}$ of Figs. 8.2 and 8.3. The nonlinear regression curve (solid) as well as two limiting slopes (dashed lines)—shifted parallel upward for better viewing—are shown within their indicated range of ε.

drimer molecule rotated into several different orientations with respect to the grid system, and under a series of differently drawn projections. Although the range of box size ε covers a factor of more than a decade, this is deceptive because efficient and truer covering requires small ε, as can be seen from the nonlinearity of the plot over the entire range of ε. Consequently, the effective range here is more likely $1.67 \leq \varepsilon \leq 4$—clearly insufficient unless a greater density of different ε within this range, fortified by a serious error analysis, is used as compensation. It is also surmised that the accentuated line thickness of the dendrimer segments in Figs. 8.2 and 8.3 (for purposes of better viewing) introduces an ambiguity; it is here that a color printer would be of great advantage. Hence, the (on first sight so simple) box counting method needs to be done very carefully and repeatedly if the result is to be trusted. It would also be wise to thoroughly check any software procedures as to their correct workings!

Let us now assume that box counting dimension $d = 0.97$ of Fig. 8.4—so close to $d = 1$—can be taken to be *strictly* $d = 1$. The dendrimer molecule is, therefore, not a fractal? But *which* fractal is it not?

To see this, we ask which topological property of the dendrimer molecule is measured by box counting (and why the anticipated answer should make us confident to state that the dimension ought to be $d = 1$). To this effect, a little reflection shows that box counting enumerates the tree trunks of the dendrimer molecule between branching points, under decreasing box length [5]. But note that there are no iterative *contraction-shift* operations (Glossary 1) generating smaller and smaller trunk lengths on finer and finer detail upon further and further branching: By constructing higher stages of the molecule, we are just adding one additional $(CH_2)_3$ segment at each of the nitrogen branching points, continuing until the increasing density of the surface groups of the molecule no longer permits growth. We are reminded of the stacked Russian dolls, an object that does not reflect increasing complexity either (see Glossary 1, *self-similar*). Clearly, then in this respect $DAB(CN)_{64}$ is indeed not a fractal and its dimension equals the topological dimension of its constituting parts, namely, line segments of dimension $d = 1$.

In fact, to turn the dendrimer into a tree fractal [6], the carbon number of the methylene chain segments connecting the nitrogen atom branching points would have to change from stage to stage—say $(CH_2)_6$, $(CH_2)_5$, $(CH_2)_4$, . . . for stages 1, 2, 3, ...

But what about a mass fractal dimension d of the dendrimer, mass scaling with distance as $M(R) \sim R^d$? An obvious approach is to (a) draw concentric circles under increasing radii R about a center within the projected dendrimer, (b) counting the number $N_R(R)$ of circle–dendrimer intersections, (c) repeating the method about different centers, and (d) plotting $\log[N_R(R)]$ versus $\log(R)$. Of course, an estimate is necessarily crude as, first, the number of line segments of $DAB(CN)_{64}$ jumps at some particular value of radius R (see Fig. 8.3) and, second, the effective range of R is less than a factor of 10. Nevertheless, I tried it about one point (the origin of a branch), obtaining slope $s = -d$ or $d \equiv d_M = 1.8 \pm 0.1$ (see Fig. 8.5).

The correct numerical values of mass–distance scaling dimension d for finite clusters of branched, loopless polymer molecules (lattice animals) are known [7].

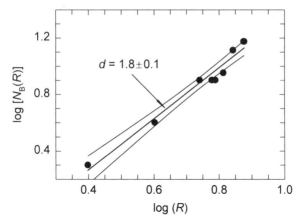

Fig. 8.5 Estimate of the mass fractal dimension of the dendrimer molecule of Fig. 8.3 from a log–log plot (decimal) of the number of intersections $N(R)$ of circles of radius R with segments of the dendrimer structure. The 95% confidence limits are also drawn.

They are $d = 1.56$ for $E = 2$ embedding space appropriate to the projection of $DAB(CN)_{64}$—my rough estimate yielding $d = 1.8$—and $d = 2$ for $E = 3$ embedding space, the actual $DAB(CN)_{64}$ molecule. Interesting: The molecule effectively is all surface.

In conclusion: Dendrimer molecule $DAB(CN)_{64}$ (embedded in $E = 3$ Euclidian space) is a one-dimensional tree fractal and a two-dimensional mass fractal. Indeed, a valid point can be made that it *has* to be a mass fractal of dimension $1 < d < 3$ because unrestricted branching is constrained by increasing steric hindrance from end groups so that mass piles up at its very surface. Dendrimer $DAB(CN)_{64}$ does not need a code for growth arrest. Its mass fractal dimension is larger than the topological dimension of the pieces it is made of ($d = 1$) but smaller than the dimension of its physical-maximal growth space ($d = 3$).

Note that mass fractal dimension $d = 2$ corresponds to that of self-avoiding random walk of a linear chain in a theta solvent (balanced attractive-repulsive interlink forces).

8.2 ON DETERMINANTS GENERATING FRACTALITY IN CHEMISTRY

The reader must have perceived that fractal dimensions are, in the majority of cases, derived from slopes of power law relations. However, I get the impression that my colleagues in chemistry do not hold the predictive relevance of log–log plots in overly high esteem [8]. Hence, in order to be convincing, experiments must reach over a yardstick range as far as the system permits—a factor of 10 has been suggested as the minimal requirement [9]—although system-dependent restrictions apply (see Section 7.6) and a priori fundamental scaling range limitations have been suggested [10]. At any rate, extensive data accumulation with subsequent averaging and objective error analysis are required.

It is useful to reiterate the conditions assuring that the exponent β of an observed power law, $P \sim R^\beta$, represents a fractal dimension. This raises the question under which conditions the power law exponent (usually called β) *is* a fractal dimension: (a) Must object or property P be self-similar, or at least, be statistically self-similar? (b) Are fractal objects best characterized as objects having dimensions that exceed the topological dimensions of their constituent parts [11]? (c) Does the property of finer and finer detail identify an object as fractal? Indeed, it seems that each of the points raised relates, in a circuitous way, to all others.

However, in applications of fractals to chemistry it serves no purpose looking for finer points of definitions or dwelling on exceptions to a rule. Assuming that experimental manipulations are carried out under the best possible conditions of total yardstick range and with appropriate error analysis [12], the application of fractal theory to the data is warranted under a few basic conditions and experimental outcomes.

On a fundamental level, we are persuaded that a *limit structure* exists. If we cannot imagine a set of transformations on the points of some initial object that would turn, say its smooth surface into a very wrinkled shape by an *iterative contraction-shift* formalism, or differently expressed, a *folding* process, there would be no sense in searching for its fractal properties.

On an observational–experimental level, the system is inhomogeneous, irregular, and its regions of interest are lacking significant translational symmetry. In addition, experimental and theoretical approaches using the "usual" classical methods of Euclidian geometry fail or become unwieldy (too many input parameters). Furthermore, numerical values of scaling slopes, ensuing from log–log plots of various system properties versus a (linear) size parameter inherent in the system's structural patterns, show the following behavior: (a) The scaling slopes depend on the particular system property measured; (b) are different from the integral exponents of normal or classical scaling behavior; (c) may undergo an abrupt change within a narrow region of the scaling range, a crossover that can be related to a simultaneous changeover of the observed system property or observational conditions; and (d) stay within the limits determined by the topological dimension of the minimally embedding space of the experiment; for example, diffusion of a flat culture in a Petri dish is embedded in $E = 2$, diffusion of a solute in a beaker is embedded in $E = 3$. Finally, (e) numerical evaluations of the scaling slopes are expected to be physically geometrically reasonable within the context of their predictions, and their values (in terms of fractal dimensions or their dependent quantities) should closely approach those of the limit structure if the latter is known or can be simulated.

Therefore, if these conditions are met, the particular object is—for all theoretical and practical purposes in chemistry—a fractal and the exponent of the scaling relation is directly related to the fractal dimension of the *particular property of the object scaled under the applied measuring approach or method.*

The questions raised above can then be answered readily. First, irregular objects met in chemistry, with dimensions exceeding the topological dimension of their constituent parts, *must* involve at least *some* degree of finer and finer detail on purely geometrical grounds. Second, because many processes in chemistry involve random aspects, appearance of statistical self-similarity or self-affinity ought to be expected or, at least, suspected: Power law behavior guarantees that *some* regions look like others [2]. Finally, self-similarity per se is not a precondition for fractality: A fractal may be self-similar, but self-similarity does not make a fractal [13].

As to the mechanics that generate fractals of interest to chemistry, two relevant examples are added to the notion that "fractality may be the result of a random-searching, iterative, optimally space-filling growth process under some constraint" (see Section 3.2.3): (a) A randomly coiling, linear polymer chain of equal monomer units swollen in a good solvent—that is, prevalence of repulsive interlink forces—cannot fill $E = 3$ space at liberty because it cannot penetrate its own segments. Hence, the ensuing self-avoiding random walk generates a fractal polymer chain—a fractal curve—of fractal dimension $d = \frac{5}{3} \approx 1.67$ [14]. The numerical value of its dimension exceeds topological dimension $d = 1$ of its constituent linear segments by $\frac{2}{3}$. (b) A randomly coiling surface in a good solvent is spatially restrained in its search for positions in $E = 3$ space by the impossibility to wander through its own surface sections. The fractal dimension of the self-avoiding random walk of such a surface is $d = \frac{7}{3} \approx 2.33$ [15]. The numerical value of its surface dimension exceeds topological dimension $d = 2$ of its constituent pieces by $\frac{1}{3}$. Indeed, $d = \frac{7}{3}$ should be about the (lower limit of the) surface dimension of a tightly crumpled ball of a large piece of thin paper.

As to fractons—the numerical quantities characterizing general flux phenomena over fractal structures that contain loops, branches, dangling bonds, and similar traps—they are obviously of immediate applicability to a large variety of dynamical processes in chemistry. A fracton dimension $\bar{d}$ gives, so to speak, details of its underlying fractal in terms of the latter's degree of structural connectivity. This information is not available from a fractal dimension but must be obtained by recording the *delay* of a probe actually "walking" over the structure of the fractal.

Experimentally, we have seen that the requirements for accurate measurements is perhaps more stringent for fractons than for fractals because of condition $1 \le \bar{d} \le 2$ (compact walk). Almost entirely, $\bar{d}$ is determined by a laboratory apparatus, computer model, and scaling argument; rarely can $\bar{d}$ be calculated by analytical procedures on the basis of the fractal's structure.

8.3 CONCLUDING REMARKS

Hopefully, this book has brought across an acceptance that fractals and fractal aspects are quite pertinent to chemistry and its applications. There is now little doubt on the scientific merits of fractals. To score another point concerning applications, I came across a reference in the polymer section of *Chemical Abstracts* to a patent specifically written to exploit the superior properties of fractal polymers as interstitial materials for adsorption and their use as additives improving ion exchange, filtration, and membrane separation [16].

To keep this book manageable to its readers (and the author), the descriptions of the numerous fractal aspects were kept concise and relatively brief. In fact, many current and projected papers of reported fractal aspects in various disciplines within chemistry, or direct and indirectly related to it, had to be left out: Autocatalytic reactions, cellular automata, DNA growth and packing, dissolution of compacted materials and of drugs, fracture, gaseous emanation from and leakage through soil, membranes, multifractals, oil recovery, optimized sampling methods after chemical spills, thin films, and so on.

The main purpose here is to offer necessary backgrounds of fractal theory and practice in chemistry, to show the generality of fractal concepts and their relevance to diverse chemical phenomena, and to encourage further and wider reading, research, and applications. There is no reason to be bewildered about the fractal as a limit structure and the fractal as an object in chemistry. One is a mathematical theory, the other a reality as seen by its concept.

NOTES AND REFERENCES

[1] D. J. Farmer, E. Ott, and J. A. Yorke, The dimension of chaotic attractors, *Physica* **7D,** 153 (1983).

[2] H.-O. Peitgen, H. Jürgens, and D. Saupe, *Fractals for the Classroom, Part One: Introduction to Fractals and Chaos* (Springer, New York, 1992), p. 240. See also Figs. 2.1 (p. 26) and 3.2 (p. 40) in K. Falconer, *Fractal Geometry* (Wiley, New York, 1990), for instructive demonstrations of Hausdorff and box counting covers of an arbitrary set.

[3] E. M. M. de Brabander-van den Berg and E. W. Meijer, Poly(propylene imine) dendrimers: large-scale synthesis by heterogeneously catalyzed hydrogenations, *Angew. Chem. Internl. Ed. Engl.* **32,** 1308 (1993).

[4] If branching continued ad infinitum, the (virtual) structure would occupy a space of infinite dimension; it is then called Bethe lattice of coordination number (nearest neighbors) z; see D. Stauffer and A. Aharony, *Introduction to Percolation Theory* (Talor & Francis, London, 1992), p. 27.

[5] See Chapter V, Section 16–17 of [2] in Section 1.1.

[6] D. S. Berger, Modification of a simple fractal tree growth scheme: Implications on growth, variation, and evolution, *J. Theor. Biol.* **152,** 513 (1991).

[7] S. Havlin, Z. V. Djordjevic, I. Majid, H. E. Stanley, and G. H. Weiss, Relation between dynamic transport properties and static topological structure for the lattice-animal model of branched polymers, *Phys. Rev. Lett.* **53,** 178 (1984).

[8] See [2] in Section 2.1.

[9] J. M. Drake, J. Klafter, and P. Levitz, Fractal Surfaces: *A Dilemma in the Characterization* of *Porous Glasses,* in *Chemically Modified Surfaces,* H. A. Mottola and J. R. Steinmetz, Eds. (Elsevier, Amsterdam, 1992). See also [6] in Section 2.2.

[10] See [41] in Section 7.6.

[11] See [2] in Section 1.1.

[12] I. R. Gatland, A weight-watcher's guide to least-squares fitting, *Comput. Phys.* **7,** 280 (1993).

[13] See [2], p. 230.

[14] See pp. 329–330 in [2] of Section 1.1.

[15] A. Maritan and A. Stella, Scaling behavior of self-avoiding random surfaces, *Phys. Rev. Lett.* **53,** 123 (1984).

[16] Chemical Abstracts, *38-Plastics Fabrication and Uses,* abstract 126: **265011f** (1997).

APPENDIX 1

TABLE OF FREQUENTLY USED DIMENSIONS

Process	Fractal Dimension	Fracton Dimension	Embedding Space
Diffusion-Limited Aggregation			
Particle–Cluster	1.68	1.26	2
Particle–Cluster	2.5	1.39	3
Cluster–Cluster	1.7–1.8	1.17	3
Reaction-Limited Aggregation			
Cluster–Cluster	1.53		2
Cluster–Cluster	1.98–2.11		3
Solvent-Swollen Linear Polymer Chain			
Prevalent repulsive interlink forces	$^5/_3 = 1.6$		≥ 2
Balanced interlink forces	2		≥ 2

Source: S. Havlin and D. Ben-Avraham, Adv. Phys. **36,** 695 (1987).

APPENDIX 2

LIST OF FUNDAMENTAL EQUATIONS

Equation Number	Section Number	Description
1.1	1.2	Scaling of number of elements $\mathcal{N}(\varepsilon)$ with linear size ε
2.1	2.2	Scaling of surface area $\mathcal{N}(\varepsilon)\varepsilon^2$ with linear size ε
2.2	2.2	Self-similar mass (M) − distance (R) scaling
2.3c	2.2	Self-affine length (L_C) − distance (R) scaling
2.4	2.3	The d-dimensional content
2.7	2.3.2	Scaling of pore surface $-dV/d\rho$ with pore radius ρ
2.10	2.3.3	Scaling of film volume $\mathcal{N}(z)z^3$ with film thickness z
2.11	2.3.3	Scaling of film surface area $\mathcal{N}(z)z^2$ with film thickness z
2.17	2.3.5	Scaling of a generalized property P with a generalized linear extent scale
3.1	3.1.1	Self-similar mass (M) − distance (R) scaling
3.5	3.1.2	Scaling of particle number $N(R)$ with distance R
3.6	3.1.2.2	Scaling of scattered radiation wave vector q with scattering angle θ
3.8	3.1.2.2	Scaling of scattered radiation with wave vector q for a purely mass fractal
3.9	3.1.2.2	Scaling of scattered radiation with wave vector q for a purely surface fractal
3.12	3.1.3	Scaling of a generalized property with distance R
3.13a	3.1.3	Scaling of a purely mass fractal with distance for mass, surface, and pore
3.13b	3.1.3	Scaling of a purely surface fractal with distance for mass, surface, and pore

APPENDIX 3

LIST OF FIGURES

Chapter 3

GLOSSARY 1

FRACTALS AND STRUCTURE

Attractor. An attractor $\mathscr{A}$ may arise as the *limit* of infinitely many iterations under some contraction-shift transformation (a *mapping*) $\mathcal{M}$ on some set $\mathscr{B}$ of points: $\lim_{n \Rightarrow \infty} \mathcal{M}(\mathscr{B}) = \mathscr{A}$, $\mathcal{M}(\mathscr{A}) = \mathscr{A}$. Note that mapping the attractor gives the attractor (because it is the limit). The initial set is called *initiator* or *axiom*, the mapping is called *generator, production,* or *rewriting* rule. A particular mapping is conveniently given the symbol C_n, where n indexes the stage of development within a series of mappings.

A production scheme is shown on the example of the triadic Cantor set. (a) We define its mapping as the <u>union</u> of <u>sets</u> of points generated by the action of two linear coordinate transformations, written $P \cup Q$, that operate on *all* coordinates x of the axiom or initiator, a line (of unit length). (b) Mapping P prescribes: Take any x of the line including endpoints, written $x \in [0,1]$, and contract it by factor 3; formally, $P\{x, x \in [0,1]\} \Rightarrow x/3$. (c) Mapping Q prescribes: Take any x, contract it by a factor of 3, then shift it by $+\frac{2}{3}$; formally, $Q\{x, x \in [0,1]\} \Rightarrow x/3 + \frac{2}{3}$. (d) Subsequently, take the union, $\cup$, of the two results.

The top of Fig. G1.1 shows the initial line, the axiom. The line below it, down-shifted for better viewing, demonstrates stage $n = 1$, that is to say, the results of $C_1 = P \cup Q\{x\}$ on all points of the axiom. For instance,

$$P(0) = 0 \quad P(1) = \tfrac{1}{3} \quad P(\tfrac{1}{4}) = \tfrac{1}{12} \qquad Q(0) = \tfrac{2}{3} \quad Q(\tfrac{1}{4}) = \tfrac{3}{4} \quad Q(1) = 1$$

Hence, $C_1 = P \cup Q\{x, x \in [0,1]\}$ yields the original line with its middle third missing.

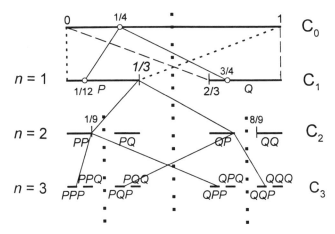

Fig. G1.1   Initiator C_0 and first three stages C_1, C_2, C_3 of the triadic Cantor set. Results of mapping $\mathcal{M} = (P \cup Q) C_0$ are shown for $x = 0$, ¼, 1. Connecting lines emerging from $x = \frac{1}{3}$ on C_1 depict contraction-shift operations $\mathcal{M}(\frac{1}{3}) = (\frac{1}{9}) \cup (\frac{7}{9})$ and $\mathcal{M}\mathcal{M}(\frac{1}{3}) \equiv \mathcal{M}^2(\frac{1}{3}) = \mathcal{M}(\frac{1}{9} \cup \frac{7}{9}) = (\frac{1}{27}) \cup (\frac{19}{27}) \cup (\frac{7}{27}) \cup (\frac{25}{27})$.

Stages C_2, C_3, and so on are generated through *iterative* mappings by $C_1 = P \cup Q$ on $x \in [0,1]$; obviously, we do not have to worry about the missing segments. For example,

$$P(P\,(1)) = PP\,(1) = P\,(\tfrac{1}{3}) = \tfrac{1}{9} \qquad Q(P(1)) = QP(1) = Q(\tfrac{1}{3}) = \tfrac{7}{9}$$

$$PPP(1) = \tfrac{1}{27} \quad PQP(1) = \tfrac{7}{27} \quad QPP(1) = \tfrac{19}{27} \quad QQP(1) = \tfrac{25}{27}$$

The rule of multiple operators is that the operator next to the argument operates first; in other words, the sequence of operations "moves to the left." Furthermore, nomenclature such as $QQ = Q^2$ does *not* imply exponentiation. For instance, $Q^2(\tfrac{1}{4}) = QQ(\tfrac{1}{4}) = Q(\tfrac{3}{4}) = \tfrac{11}{12} \neq [Q(\tfrac{1}{4})]^2 = \tfrac{9}{16}$.

By discarding forever the middle thirds of ever shrinking line segments by contraction-shift operations $P \cup Q$, is there anything left? There is. Consider that each operation by $P \cup Q$ generates endpoints. However, points other than endpoints remain, which can be demonstrated by writing x in a triadic base

$$x(\text{triadic}) \equiv x_3 = a_1(3^{-1}) + a_2(3^{-2}) + \cdots a_n(3^{-n}) + \cdots \text{ with } \quad a_n = 0, 1, 2$$

For instance, decimal fraction $\tfrac{1}{3} = (\tfrac{1}{3})_{10} = 0.\overline{3}_{10}$ (the overstrike signifying forever repeating digits) reads $0.1\overline{0}_3$ or simply 0.1_3 in triadic. The advantage is that division by 3 in a triadic representation of numbers is represented by a left shift of the point by one place. Thereby, contraction operations by a factor of 3 are readily enumerated. As it turns out, (a) triadic expansions of points containing digit 1 are in the empty spaces of the stages; for instance, $0.5_{10} =$ $0.1111 \cdots 11 \cdots {}_3 = 0.1\bar{3}$. (b) Triadic expressions of endpoints have infinitely repeating digits 2 or 0, for instance, $(\frac{1}{3})_{10} = 0.1_3 = 0.022 \cdots 22 \cdots {}_3 =$ $0.\bar{2}_3$, $(0.0)_{10} = 0.0\bar{3}$. (c) This leaves one more possibility, namely, all infinitely long sequences of random numbers containing digits 2 and 0, such as $0.022002220202_3 \cdots$, which you can write down until the end of time, never finding two that are equal [1].

Infinitely many points of the original Euclidian object are cut out. In fact, the series development adding the discarded middle thirds yields precisely the original length: lim $[\frac{1}{3} + 2(\frac{1}{9}) + \cdots 2^{n-1}/3^n \cdots + \cdots] = 1$, for $n \Rightarrow \infty$. Yet, infinitely many points remain, forming a mathematical limit object or attractor that is not contained in its stages C_n—as limit lim $= 1$ is not contained in the stages $2^{n-1}/3^n$ of its series.

The triadic Cantor set is the *intersection* of all $(P \cup Q)^n \{x, x \in [0, 1]\}$,

$$\mathscr{A}(\text{Cantor}) = \cap_{n=0}^{\infty} (P \cup Q)^n \{x, x \in [0,1]\}$$

The question coming to mind is whether such a strange fractal object could possibly have *any* reference to chemistry? Certainly. For instance, the relatively intense spectral lines of some polyatomic molecule represent, upon increasing spectral resolution, successive stages of a Cantor-like set, from its high-frequency electronic transitions (few) of the order of 10^{15} s^{-1}, to vibration–rotation levels (many more) of the order of 10^{13} s^{-1}, and on down to many lines of hyperfine structures on the scale of a few 10^6 s^{-1} from nuclear spin quadrupole–rotation interactions, and so on [2].

Contraction-shift operation. See under **Attractor.**

Covering, measure, Hausdorff p-dimensional measure. We have used a measure or covering in Fig. 1.3 to illustrate 'dimension'. Notice that the covering discussed there was of the nature of self-similar patterns; in other words, the pieces into which the overall object was divided had the exact shape of the object— only correspondingly smaller. Such a type of covering may be too restrictive when measuring irregular objects that are not self-similar.

A more general covering is the Hausdorff measure of a (compact) subset $\mathscr{A}$ of a Euclidian metric space, where $\mathscr{A}$, in turn, is the union of subsets A_i, $\mathscr{A} = \cup_{i=1}^{i=\infty} A_i$ [3]. In principle, there is no prior knowledge of the dimension of $\mathscr{A}$. First, a diameter of A_i, diam(A_i), is defined as the *supremum*, or sup $= smallest$ *upper bound,* of the distance between any two points x, y in A_i: diam$(A_i) =$ sup$\{|x\text{-}y|; x, y \in A_i\}$. This definition accommodates the most distant points apart, but no more. Second, sets A_i are covered by balls of volume $[\text{diam}(A_i)]^p$, leaving parameter p, $0 \leq p < \infty$, to be determined. (Consider, e.g., that measuring a two-dimensional surface, we would reasonably guess $p = 2$. Hence, the balls would turn into disks.) Third, we sum up the covering p-dimensional volumes over sets A_i, $\sum_{i=1}^{i=\infty} [\text{diam}(A_i)]^p$, striving for the most economical measure, the *infimum* or inf $= largest$ *lower bound.* (Replacing the union of A_i, $\cup_i A_i$, by

$\sum_i A_i$ is allowed if the *intersection* of the A_i is zero or small: sets A_i do not overlap too much.)

To get the desired infimum, an upper bound η is imposed on $\text{diam}(A_i)$, $\text{diam}(A_i) < \eta$, $\eta \Rightarrow 0$. This procedure leads indeed to the most economical measure: Overlap of the $[\text{diam}(A_i)]^p$ is minimized as points in the sets A_i are eventually not counted repeatedly. With the ever shrinking $\text{diam}(A_i)$, quantity inf $\sum_i [\text{diam}(A_i)]^p$ can only increase or stay the same—an outcome shown by calculation. Thus, under $\eta \Rightarrow 0$, the inf of the p-dimensional covering of subsets A_i approaches a limit, which is the so-called Hausdorff p-dimensional measure $M(\mathcal{A}, p)$ of $\mathcal{A}$,

$$M(\mathcal{A}, p) = \lim_{\eta \Rightarrow 0} \inf \{ \textstyle\sum_{i=1}^{i=\infty} [\text{diam}(A_i)]^p, \text{diam}(A_i) < \eta \}$$

$M(\mathcal{A}, p)$ can assume three values: (a) $M = 0$ if $p > d_H$, (b) $M = \infty$ if $p < d_H$, and (c) $M = $ finite (or $M = 1$ if normalized) if $p = d_H$; d_H is called *Hausdorff dimension*.

Example 1: (a) Imagine you would have to take the measure of a 8×11 page of paper by using lines, $p = 1$, as covering balls. It would have to be *a very thin pencil* and would take forever to cover the page: $M = \infty$ because $p < d_H = 2$. (b) Now consider taking the measure of the paper with *volume* elements made of clay, $p = 3$, as covering balls. To get the page area you would have to squish the clay cover down flat into 'equatorial planes'. Now $M = 0$ because $p > 2$.

Example 2: The fractal object $\mathcal{A}$ is the triadic Cantor set (see *attractor* above). We cover its pieces by line segments and economical covering is feasible using one segment of unit length for $n = 0$ (initiator), two segments of length $\frac{1}{3}$ each for stage $n = 1, \ldots$, and 2^n segments of length $(\frac{1}{3})^n$ each for stage n. The infimum (largest lower bound) of all coverings is therefore

$$\inf \textstyle\sum_i [\text{diam}(A_i)]^p = \inf \sum_n [(\frac{1}{3})^n]^p = \inf 2^n [(\frac{1}{3})^n]^p \qquad n = 1 \text{ to } \infty$$

Condition $(\frac{1}{3})^n < \eta$, $\eta \Rightarrow 0$, is equivalent to $n \Rightarrow \infty$, hence, $M(\mathcal{A}, p) \leq \lim_{n \Rightarrow \infty} 2^n (\frac{1}{3})^{np}$. We try $p = 1$ and get $M(\mathcal{A}, 1) = 0$: Choice $p = 1$ is *too large*. We try $p = 0$ and get $M(\mathcal{A}, 0) = \infty$: Choice $p = 0$ is *too small*. However, $p = \log 2/\log 3$ yields $M(\mathcal{A}, p) \leq 1$. Hence, dimension $d_H(\mathcal{A}) = \log 2/\log 3 \approx 0.63$ is a good proposition.

For writing the opposite inequality, $M(\mathcal{A}, \log 3/\log 2) \geq 1$ and, consequently, proving that $M(\mathcal{A}, \log 3/\log 2) = 1$, see [4].

Dimension. See under **Self-affine fractal, self-similar dimension.**

Intersection, set, union. It is helpful to briefly define these concepts; they are frequently used in the theoretical literature on fractals. For example, a symphony orchestra is a set of musicians. The string section, brass section, wind section, percussion section, and the conductor are each a *subset, M_S*, of the orchestra M: It is their union $\cup_S M_S = M$, which makes the music. What is a subset? Patsy, the piccolo flute player, is a member of the set wind section M_W, but is also a member of the set orchestra M. Therefore, M_W is a subset of M, $M_W \subset M$.

A subset may have its own subsets: The violins, violas, cellos, bass players are subsets of subset string section. Usually, the intersection ∩ of the subset of musicians, M_S, and the subset of the maestro, M_C, is strictly empty: $M_S \cap M_C = \varnothing$. But $M_S \cap M_C \neq \varnothing$ when the conductor conducts and is the soloist.

The union of the sets A and B, $A \cup B$, is the set whose members are those objects x such that x belongs, at least, to one of the two sets A, B; that is to say $x \in A$ or $x \in B$. Note that "or" is the logical connective or, used in the inclusive sense: For example, assume that P or Q are true, which allows (a) P true, Q false, (b) P false, Q true, (c) P true, Q true. If P and Q are false; then only P false, Q false are admitted.

The intersection of sets A and B, $A \cap B$, are those objects x such that x belongs to set A *and* to set B [5].

Metric. How would we know that a limit object, a fractal, is indeed approached by iterative contraction-shift operations? A useful method would be exploring whether a distance measure or metric, m, gets ever closer to the limit under ever-increasing number of iterations. But distance between what? It is easy to define a metric between two points a, b of some space. The conditions of a metric are (a) Distance $a \Rightarrow b$ ought to equal distance $b \Rightarrow a$; formally, $m(a, b) = m(b, a)$. (b) $m(a, b) \geq 0$, with equality—$m(a, b) = 0$—only if $a = b$. (c) $m(a, c) + m(c, b) \geq m(a, b)$. Examples: Assign x, y, z coordinates to points a and b, $a = \{a_x, a_y, a_z\}$ and $b = \{b_x, b_y, b_z\}$. (a) The usual *Euclidean metric* (helicopter ride) is defined as $m(a, b) = [(a_x - b_x)^2 + (a_y - b_y)^2 + (a_z - b_z)^2]^{1/2}$. (b) The lesser familiar *Manhattan* metric (taxi ride with subsequent ascent by elevator) is defined by $m(a, b) = |a_x - b_x| + |a_y - b_y| + |a_z - b_z|$, where $|u|$ signifies the absolute value (amount) of u. (c) Inequality relation $m(a, c) + m(c, b) \geq m(a, b)$ is readily understood by watching a squirrel traversing a lawn with a few isolated trees on it (the corresponding graph is drawn when snow has fallen).

However, a fractal and its various finite iterations (or stages) are sets of points; herein lies the difficulty of defining a distance between two stages, say n and $n + 1$. For example, what is the distance between set Bronx (B) and set Suffolk County (S)? Going from the Yankee Stadium (Yankee Stadium $\in B$) to Montauk Point ($\in S$) is not the same distance as going from Brookhaven National Laboratory ($\in S$) to the Bronx Zoo. The method for obtaining a metric between sets A and B is demonstrated in Fig. G1.2. First, the perimeter of set A, A_δ, is "puffed up"—keeping its shape intact (collar of A)—to just include set B. Second, a perimeter around set B, B_δ (collar of B), is enlarged in the same manner to just include set A. Third, the larger of the two collars A_δ, B_δ of sets A, B is taken as metric, called *Hausdorff distance* $h(A, B)$ between sets A and B. This definition clearly avoids violation of rule $m(A, B) = m(B, A)$ and satisfies the remaining conditions of a metric (see above) [6]. (Sets A and B are compact; their points do not wander off.)

Scaling. A general method for predicting behavior of a dependent quantity under a

$$h(A, B) = \inf\{\delta : A \subset B_\delta \text{ and } B \subset A_\delta\}$$

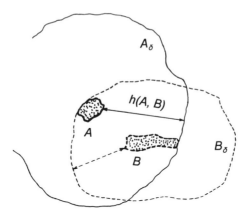

Fig. G1.2 Sketch of the method for obtaining the Hausdorff distance $h(A, B)$ between compact sets A and B. Shown are perimeter A_δ of set A and equidistant to it, as well as perimeter B_δ of set B and equidistant to it, drawn such that the perimeter around one set just encloses the other set; each set has, so to speak, ballooned up to fully but economically include the other.

scale change of an independent variable, with retention of the initial functional relationship. In other words, the scaled function equals its initial form multiplied by a constant, function-independent factor. For instance, volume V of a sphere of radius r, $V(r) = (4/3)\pi r^3$, if scaled under $r \Rightarrow br$, changes to $V(br) = (4/3)\pi(br)^3 = b^3(4/3)\pi r^3 = b^3 V(r)$; hence $V(r) = b^{-3}V(br)$. This is an example of an *affine* scaling; that is, multiplication of the independent variable by constant factor (here b) requires multiplication of the dependent variable by a different constant factor, here b^{-3}. (See also under *self-affine* fractal.) If both scaling factors are equal, the scaling is said to be *self-similar* (see under this entry).

Self-affine fractal. A self-affine fractal can be described as the union of parts each of which is a copy of the whole object but scaled by different factors in different coordinate directions. The coordinate transformations yielding the self-affine object are readily written in terms of matrices with different constant multipliers (scaling factors) a, b, such as $\begin{vmatrix} a & 0 \\ 0 & b \end{vmatrix}$. Self-affine systems have two different fractal dimensions, one at microscopic (*local*) scales from down-sizing and one at macroscopic (*global*) length scales from up-sizing. They obey $d_{local} > d_{global}$.

   As an example, we take as initiator rectangular area $a \times b$, a = height = 2, b = width = 3. It helps to take a piece of paper, pencil, and eraser while referring to Fig. G1.3. Down-generator: (a) Partition the initiator $a \times b$ into 6 equal, self-affine rectangular areas scaled by factor $1/a = \frac{1}{2}$ on the height, $1/b = \frac{1}{3}$ on

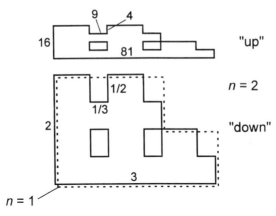

Fig. G1.3 Graphical demonstration of down- and up-scaling of a self-affine object. Starting with a rectangular initiator of height 2 and width 3, stage $n = 1$ (dotted lines) and stage $n = 2$ (solid) under down-scaling are shown slightly offset for better viewing. The upper fractal depicts stage $n = 2$ for up-scaling. Objects are arbitrarily adjusted to identical width.

the width; hence, 6 1×1 pieces. (b) Now discard the upper right 1×1 piece; the result is rendered in Fig. G1.3 by its dotted-line frame, slightly shifted off for better viewing. (c) Continue in the same way, namely, partition each remaining rectangular 1×1 piece into 6 each scaled by factor $1/a$ on its height and by factor $1/b$ on its width (steps a), and subsequently discard the upper right piece in all of them (steps b). These operations yield the second stage, $n = 2$, shown in Fig. G1.3 by solid lines. Note how empty area segments arise, within and at upper and right edges of the object.

And so on: Area A of stage $n = 1$ amounts to $A(n = 1) = 5 \times (1 \times 1) = 5 \times [2 \times (\tfrac{1}{2})] \times [3 \times (\tfrac{1}{3})]$, area $A(n = 2) = 25 \times (\tfrac{1}{2}) \times (\tfrac{1}{3}) = 25 \times 2 \times 3 \times (\tfrac{1}{2})^2 \times (\tfrac{1}{3})^2$—hence, area A $(n;$ down$) = 5^{\zeta + n}(\tfrac{1}{2})^n(\tfrac{1}{3})^n$. Exponent ζ takes care of constant product 2×3, its factors chosen quite arbitrarily here. The empty (missing) rectangles become more *vertical* with increasing n.

Up-generator: Stage $n = 2$ is chosen and scaled up to 2^ζ on height, 3^ζ on width; specifically $\zeta = 4$, as demonstrated by the object in the upper part of Fig. G1.3. It comprises 25 rectangles of area $2^2 3^2$ each, or $A(n = 2) = 5^2 2^2 3^2$. Hence, area A $(n;$ up$) = 5^{\zeta - n}2^n3^n$. The empty rectangles become more *horizontal* with increasing n.

To get the dimensions, we use Eq. (1.1), $\mathcal{N}(\varepsilon) \sim \varepsilon^{-d}$. Do we pick width factor $1/a$ or height factor $1/b$ as the linear yardstick ε? It would seem that for down-scaling (increasingly vertical pieces) we ought to take the smaller of the two, $\varepsilon_{n(b)} = (1/b)^n = (\tfrac{1}{3})^n$. For up-scaling (increasingly horizontal holes), matters are reversed: $\varepsilon_{n(a)} = a^n = 2^n$. Noting that $\mathcal{N}(\varepsilon_n) = A(n)/\varepsilon_n^2 \sim \varepsilon_n^{-d}$, we then get

$$\mathcal{N}(\varepsilon_n; \text{down}) = 5^{n + \zeta}(\tfrac{1}{2})^n(\tfrac{1}{3})^n/(\tfrac{1}{3})^{2n} \sim [(\tfrac{1}{3})^n]^{-d} \quad \varepsilon_n = (\tfrac{1}{3})^n$$

$$\mathcal{N}(\varepsilon_n; \text{up}) = 5^{\zeta - n}2^n3^n/2^{2n} \sim [(2)^n]^{-d} \quad \varepsilon_n = 2^n$$

Taking logarithms on both sides and thereafter $\lim \varepsilon \Rightarrow 0 = \lim n \Rightarrow \infty$, leads to dimension $d_{local} = 1 + \log(\frac{5}{2})/\log 3 = 1.83 \cdots$ under down-scaling and to $d_{global} = 1 + \log(\frac{5}{3})/\log 2 = 1.74 \cdots$ under up-scaling. Indeed, $d_{local} > d_{global}$.

Down-scaling may be arrested by a large lattice constant, up-scaling may be voided by insufficient size of the object [7].

Self-similar. A set of points P is called self-similar if it is possible to find a finite number S of subsets $p_S \subset P$, with $\cup p_S = P$, such that for all S the relation $P = \alpha_S(p_S)$ is valid; α_S is the scaling operation. [Recall that $\subset$ is the usual set inclusion symbol and $\cup$ signifies union of.] In words; P can be considered the union of parts each of which looks like P, except for absolute size.

Examples: (a) A set of Russian nesting dolls, матрёшка. I do not think it is a fractal: There is no generation of finer and finer detail or greater and greater complexity; zooming-in on the limit would merely show the infinitely small doll. This outcome gives a warning: Self-similarity does not necessarily engender fractality. (b) The triadic Cantor set (see Fig. G1.1): The generator C_1 consists of two copies of the initiator scaled by $\frac{1}{3}$ and separated by $\frac{1}{3}$; in turn, stage C_2 consists of two copies of C_1 scaled by $\frac{1}{3}$ and separated by $\frac{1}{3}$, and so on. Self-similar scaling is the simplification of self-affine scaling, as both scaling parameters are identical.

Self-similarity dimension. The value of the ratio of the logarithms of $\mathcal{N}_n(\varepsilon)$, the number of self-similar pieces to cover (*measure*) the nth contraction stage of the object, and $1/\varepsilon$, the reciprocal value of the linear contraction factor ε from the $(n-1)$th contraction to the nth contraction stage. Using $1/\varepsilon$ instead of ε leads to positive slopes—see Eq. (1.1). The resulting dimension may be checked by reapplying the method to another pair of values of $n-1, n$. Example: The triadic Cantor set (Fig. G1.1). (a) Take stage C_2, with $\mathcal{N}_2(\frac{1}{9}) = 4$, therefore $d = \log [\mathcal{N}_2(\frac{1}{9})]/\log [1/(\frac{1}{9})] = \log 4/\log 9 = \log 2/\log 3 = 0.63$. (b) Take stage C_n, with $\mathcal{N}_n(\varepsilon) = 2^n$, $\varepsilon = (\frac{1}{3})^n$, thus $d = \log (2^n)/\log [1/(\frac{1}{3})^n] = \log 2/\log 3$.

Topological dimension. The values of a topological dimension can only be integral, 1, 2, 3, and 0. Dimension 0 is that of a point. Dimension 1 is that of a space bordered by two points (imagine a line cut at two, noncoinciding locations). Dimension 2 is that of a space limited by segments of topological dimension 1 (and analogously for dimension 3).

It will be seen on all examples given that a fractal dimension *exceeds* the topological dimension of the pieces of which it is made of [8].

Zero set. The zero set of amplitude function $Y(X)$ is defined as the set of points (instants) X where $Y(X) = 0$ in the appropriate record of Y. For example, (a) the zero set of part (*c*) of Fig. 1.2 is $\{0, 2, 4, 6, 12, 14\}$. (b) The zero set of a horizontal line drawn through Fig. 1.1(*a*) is a random Cantor set, consisting of all intersections with the Y–X profile.

NOTES AND REFERENCES

[1] H.-O. Peitgen, H. Jürgens, and D. Saupe, *Chaos and Fractals* (Springer, New York, 1992), p. 70.

[2] W. G. Harter and C. W. Patterson, Theory of hyperfine and superfine levels in symmetric polyatomic molecules. Trigonal and tetragonal molecules: Elementary spin-½ cases, *Phys. Rev. A* **19,** 2277 (1979).

[3] M. F. Barnsley, *Fractals Everywhere* (Academic, New York, 1993), p. 195.

[4] K. J. Falconer, *The geometry of fractal sets* (Cambridge University Press, Cambridge, UK, 1985), p. 15.

[5] B. Mendelson, *Introduction to Topology* (Dover, New York, 1990).

[6] See [1], pp. 289–290.

[7] P. Pfeifer and M. Obert, Fractals: *Basic Concepts and Terminology,* in *The Fractal Approach to Heterogeneous Chemistry,* D. Avnir, Ed. (Wiley, Chichester, UK, 1989), pp. 34–36.

[8] B. B. Mandelbrot, *The Fractal Geometry of Nature* (Freeman, New York, 1983), p. 361. For a source on the principles of topology, see M. A. Armstrong, *Basic Topology: Undergraduate Texts in Mathematics* (Springer, New York, 1983).

GLOSSARY 2

FRACTALS AND DYNAMICS

Anomalous diffusion. We call a transport process anomalous when its mean-squared Euclidian distance $\langle R^2 \rangle$ traversed during time interval t is shorter (takes longer) than predicted by the number of steps performed, $j(t) \sim t$. To account for this mathematically, Einstein's relation for regular diffusion—connecting the average squared distance traversed $\langle R^2 \rangle$, diffusion constant K_0 (cm²/s), Euclidian dimension E, and time t elapsed—is modified from $\langle R^2 \rangle = 2EK_0 t \sim t = t^{2/2}$ to $\langle R^2 \rangle \sim t^{2/D_w}$, with $D_w = 2 \Rightarrow D_w > 2$. D_w is conveniently decomposed into the sum $D_w = 2 + \eta$, $\eta > 0$. The mechanics underlying these changes are easy to grasp: The random walker is delayed upon traversing an extensively looped or branched fractal structure or having to extricate itself from its dangling bonds.

To describe anomalous diffusion, an additional numerical parameter, called *fracton*, is required to characterize the connectivity of such structural-delaying elements within the fractal. Fractals of the same dimension therefore need not have the same fracton dimension.

Brillouin scattering is the Bragg reflection of light from a thermal sound wave [1]. The Bragg relation is derived from the conservation of energy and of momentum during the interaction of the *photons* of the incident laser light beam with the *phonons* (thermal motions) of the irradiated substrate. Momentum conservation yields $\hbar \mathbf{q}_{phot} + \hbar \mathbf{q}_{phon} = \hbar \mathbf{q}'_{phot}$ or simply $\mathbf{q}'_{phot} - \mathbf{q}_{phot} = \mathbf{q}_{phon}$, primed quantities pertaining to the scattered photons. From these wave vector relations we calculate wave vector $|\mathbf{q}_{phon}| = q_{phon}$ of the scattered phonon, denoting the absolute value $|\mathbf{b}|$ of vector $\mathbf{b}$ by b, as is customary. Because the Brillouin measurements involve very low frequencies, set first $\mathbf{q}_{phon} \approx 0$, hence find $\mathbf{q}'_{phot} \approx \mathbf{q}_{phot}$.

Second, evaluate $q_{phon}{}^2 = (\grave{\mathbf{q}}_{phot} - \mathbf{q}_{phot})^2$ by the rules of scalar vector multiplication $\mathbf{a} \cdot \mathbf{b} = ab\cos(\mathbf{a,b})$, abbreviating the angle between vectors $\mathbf{a,b}$ by θ, and find $q_{phon}{}^2 = 2q_{phot}{}^2\,[1 - \cos(\theta)]$ under the above momentum transfer rule $\grave{\mathbf{q}}_{phot} \approx \mathbf{q}_{phot}$. Now recall trigonometric relation $1 - \cos(\theta) = 2\sin^2(\theta/2)$ and obtain Bragg condition $q_{phon} = \pm 2q_{phot}\,\sin(\theta/2)$.

From well-known principles, we get (a) $q_{phon} = \omega_{phon}/v$, where v is the velocity of sound in the substrate, (b) $q_{phot} = n\omega_{phot}/c$, with $n =$ index of refraction of the substrate and $c =$ velocity of light, and (c) $\omega_{phon} = \omega_L - \grave{\omega}_L \equiv \Delta\omega_L$, the energy conservation law between incident ($\omega_L = \omega_{phot}$) and scattered ($\grave{\omega}_L = \grave{\omega}_{phot}$) laser radiation. Inserting (a)–(c) into the Bragg condition (see above) yields the relation between laser frequency shift $\Delta\omega_L$, incident laser frequency ω_L, velocity ratio v/c, and scattering angle θ, as $\Delta\omega_L = \pm 2\omega_L(nv/c)\sin(\theta/2)$. With a 488 nm laser, $n \approx 1$ and $v/c \approx 10^{-4}$ (water), this gives a maximal ($\theta = 180°$) frequency shift of $\pm \Delta\omega_L \approx 4$ cm^{-1} in the far-infrared. Instrumentally, the scattered light is analyzed by a high-resolution Fabry–Perot interferometer [2].

Compact, noncompact, recurrent, nonrecurrent walk. As an illustration, the probability $p(t,0)$ of returning to the origin of random walk, $p(t,0) \sim t^{-\bar{d}/2}$, is computed for fracton dimension (a) $\bar{d} = 1.6 < 2$ (compact) and (b) $\bar{d} = 2.3 > 2$ (nonrecurrent walk) at three common time intervals $t = 3, 10, 100$. For (a) we get $p(3,0) = 0.42$, $p(10,0) = 0.16$, $p(100,0) = 0.03$. For (b) we obtain probabilities 0.28, 0.07, 0.005. Clearly, in the case of nonrecurrent walk, the probability of the walker's return is very small at long t.

Consider further that for any space the elapsed marching time of the random walker is ultimately given by the number of its steps performed, $j(t) \sim t$: The walker cannot outrun itself. Second, take the number of *different* (lattice) sites visited during a random walk, often written $S(t)$. It is given by $S(t) \sim [p(t,0)]^{-1} \sim t^{\alpha/2}$, with $\alpha = E$ or $\alpha = \bar{d}$, accordingly. Why would the volume explored during the random walk involve quantity $S(t)$? Because in order to relate this measure correctly to the volume–distance scaling on the fractal itself, we must not count visited (lattice) sites more than *once*. Volume–distance scaling of the fractal is formulated by $S(t) \sim (\langle R^2\rangle^{1/2})^d$ which, recalling $(\langle R^2\rangle^{1/2})^d \sim t^{d/D_w} \sim t^{\bar{d}/2}$ [Eq. (4.5)], indeed proves assertion $S(t) \sim t^{\bar{d}/2}$ [3].

Next, we compute the average time t_S a random walker spends on a (lattice) site during its march: It is given by $j(t)/S(t) = t/S(t) \sim t^{1 - \alpha/2}$. Clearly, for $\alpha < 2$ residence time t_S *increases* with increasing time of the random walk. Furthermore, as the average time interval between two succeeding random jumps is constant, increasing values of t_S signify that the walker *revisits* sites as time goes on. Contrarily, condition $\alpha > 2$ implies that t_S *decreases* with increasing times: The walker may not visit all sites that could have been visited. We therefore conclude: (a) $E < 2$ or $\bar{d} < 2$ describes compact or recurrent walk; the walker frequently comes back to sites already seen. Diffusion is anomalous for walk on appropriate fractal substrates and we must use $\alpha = \bar{d}$, $D_w > 2$ [see Eq. (4.5)], and will find that $\bar{d} < d$. (b) Condition $E > 2$ or $\bar{d} > 2$; the walk is noncompact or nonrecurrent. Diffusion is regular and we must use $\alpha = 2$ or $D_w = 2$.

It is instructive to give the enumerated expressions for $S(t)$ for regular walk in

Euclidian spaces, expressed in terms of S_n = mean number of S different sites visited after n steps. They are $S_n(E = 1) \sim n^{1/2}$, $S_n(E = 2) \sim n/\ln(n)$, and $S_n(E = 3) \sim n$. For instance, for $n = 100$ random steps, (a) 10 different sites are visited and each visited site is seen 10 times in walk-space $E = 1$, (b) 22 different sites are visited, each about 5 times, in walk-space $E = 2$, and (c) each site is visited about once in walk-space $E = 3$, on average [4]. (So, remember at least the floor of the parking structure.)

To summarize the important points: Under compact diffusion, function $S(t)$ of different sites visited or volume explored during time t requires condition $E < 2$ or $\bar{d} < 2$; we must use $S(t) \sim t^\beta$ with $\beta = E/2$ or $\bar{d}/2$, respectively. On the other hand, if either dimension exceeds value 2, the walk becomes nonrecurrent; we must use $S(t) \sim t \sim j(t)$. Therefore, relation $S(t)$ (= number of sites visited for the first time) $\sim [p(t,0)]^{-1}$ (= volume explored during the random walk by Eq. (4.4)) holds only for compact or recurrent walk.

Density of vibrational states. The density of vibrational states is the number distribution $\rho(\omega)$ of all frequencies ω per unit frequency interval, $\rho(\omega) = \sum_n \delta(\omega - \omega_n) \sim \omega^{\alpha-1}$, where δ is a delta function. For long-wavelength acoustic modes in (a) Euclidian spaces of dimension E, exponent $\alpha = E$, (b) on fractal substates of dimension d and exhibiting anomalous fluxes, $\alpha = \bar{d}$ (*see spectral dimension,* below).

Extended or stretched exponential. The stretched exponential is used to describe relaxation phenomena within inhomogeneous media, including fractals. Originally, the stretched exponential function $f(t,\gamma,\tau_c) = \exp[-(t/\tau_c)^\gamma]$, $0 < \gamma \le 1$, was conceived as an empirical relaxation function (called Kohlrausch–Williams–Watts function) to fit relaxation data that could not be properly described by a single relaxation time τ of exponential behavior, $\exp(-t/\tau)$ (see, e.g., [25] in Section 4.2.4). The stretched exponential was interpreted to reflect many concurrent, independent, parallel-occurring Poisson relaxation processes $\exp(-t/\tau)$, each under its own relaxation time τ distributed by a probability distribution function $g(\gamma,\tau)$ [5]. The resulting expression, namely, $\exp[-(t/\tau_c)^\gamma] = \int g(\gamma,\tau) \exp(-t/\tau)d\tau$, explains well a wide variety of vibrational relaxation data of liquid-crystal fluids in terms of simple probabilistic concepts of a renewal process [6].

Some years back, the stretched exponential was upgraded to a universal, theoretically fundamental relation to describe manifestations of a variety of relaxation pathways [7]. In addition, the concept of *fractal time* was coined—"to describe highly intermittent self-similar temporal behavior that does not possess a characteristic time scale [8]." Presently, nothing exciting is seen in published papers that would render the concept of fractal time to be of interest and applicability to chemists. Incidentally, the stretched exponential is related to the Weibull distribution of 'first failure times' of a system with a large number of components of random failure (weakest-link principle) [9].

Fractional Brownian motion. The principle modifies random walk on Euclidian spaces by introducing a bias into the probability of the outcome of succeeding events, thereby generating a persisting memory between the states of the random walker. Recall, first, Einstein's relation for diffusion, $\langle R(t)^2 \rangle \sim t$ [see Eq. (4.1) with $D_W = 2$]. Second, rename random variable $R(t)$ to random variable $X(t)$ in space and time with respect to arbitrary origin $X(t_0)$, often set to zero, and write the variance of $X(t) - X(t_0)$,

$$\langle [X(t) - X(t_0)]^2 \rangle \sim |t - t_0|^{2H} \qquad H = \tfrac{1}{2}$$

Third, generalize the variance by allowing range $0 < H \le 1$, introducing "less randomness"—as we shall see. The approach is named fractional Brownian motion and expressed as

$$X(t) - X(t_0) = \eta_G |t - t_0|^H \equiv B_H(t) \qquad 0 < H \le 1$$

where η_G is a Gaussian-distributed random number of no interest to us here [10]. Note that $B_H(t)$ is self-affine: $B_H(t) = b^{-H} B_H(bt)$. To see how the attribute "less random" manifests itself, we compute the autocorrelation of $B_H(t) - B_H(t_0)$, taken conveniently at time point t and earlier time $t_0 - (-t)$, all averaged over the ensemble:

$$\langle [B_H(t_0) - B_H(-t)][B_H(t) - B_H(t_0)] \rangle \equiv G(t)$$

Setting now $B_H(t_0) = 0$, we need to solve autocorrelation function $G(t) = \langle -B_H(-t)B_H(t) \rangle$. We use the trick of first writing the variance of $-B_H(-t) + B_H(t)$:

$$\langle [-B_H(-t) + B_H(t)]^2 \rangle = \langle [B_H(-t)]^2 \rangle + 2\langle -B_H(-t)B_H(t) \rangle + \langle [B_H(t)]^2 \rangle$$
$$= 2[\langle -B_H(-t)B_H(t) \rangle + \langle [B_H(t)]^2 \rangle] = 2[G(t) + \langle [B_H(t)]^2 \rangle]$$

noting that $\langle [B_H(-t)]^2 \rangle = \langle [B_H(t)]^2 \rangle$. Normalization (division by $\langle [B_H(t)]^2 \rangle$) yields

$$\langle -B_H(-t)B_H(t) \rangle / \langle [B_H(t)]^2 \rangle \equiv \hat{G}(t) = (\tfrac{1}{2})\langle [-B_H(-t) + B_H(t)]^2 \rangle / \langle [B_H(t)]^2 \rangle - 1$$

Assuming the ensemble to be in a stationary equilibrium, we need not be concerned about absolute times and, thus, can shift t by $+t$, which modifies variance $\langle [-B_H(-t) + B_H(t)]^2 \rangle$ to $\langle [B_H(2t)]^2 \rangle$. Therefore, normalized autocorrelation function $\hat{G}(t)$, recalling the definition $B_H(t) \sim |t|^H$ given above, reads

$$\hat{G}(t) = (\tfrac{1}{2})\langle [B_H(2t)]^2 \rangle / \langle [B_H(t)]^2 \rangle - 1$$

$$= (\tfrac{1}{2})[(2t)^{2H}/t^{2H}] - 1 = 2^{2H}/2 - 1 = 2^{2H-1} - 1$$

Function $\hat{G}(t)$ covers three limiting situations: (a) $H = \tfrac{1}{2}$, which gives $\hat{G}(t) = 0$ for all t; there is no correlation, hence there is no trend in succeeding events: The process is completely random. (b) $\frac{1}{2} < H \le 1$ yields $\hat{G}(t) > 0$ for all t; the system shows a positive bias, implying that a trend in the random events at some time t is maintained to later times t: The system states are said to be persistent (a lucky—or losing—streak in Vegas). (c) Condition $0 < H \le \frac{1}{2}$ leads to $\hat{G}(t) < 0$ for all t; the system shows a negative bias, implying that a trend of random events at some time $t + \tau$ is opposite to that at earlier time t: The system trends are said to be antipersistent (the serious jogger at the red traffic light).

Of course, on first sight there is something odd about this because actual system fluctuations do not possess perpetual autocorrelations. However, this can be circumvented by proposing that a break in the pattern is always conceivable at some point in time by an *outside* interference (the Feds change the discount rate).

Localized phonon. First, as a reminder, a phonon is a quantized vibrational standing wave motion extending over the entire size of a crystal. Take, for the sake of demonstration, a linear chain of equidistant (lattice parameter a) points of mass m, connected by a harmonic force constant, and interacting with a periodic radiation field of angular frequency ω. The ensuing displacement coordinate u (from its equilibrium position) in a direction along the chain is given by $u_n(t) \sim \exp[i(qan - \omega t)]$, with $i = (-1)^{1/2}$, for each mass point n (longitudinal modes) if only nearest-neighbor interactions are considered. Because of a periodicity by 2π, meaning that one wavelength λ fits between mass point positions an and $a(n + 1)$, we require that factor q obeys $q(na + \lambda) - \omega t = qan + 2\pi - \omega t$, which gives $q = 2\pi/\lambda$. Quantity q is called wave vector (note that q is the phonon wave vector).

If the translational symmetry of the crystal is lost, as in a disordered material, the phonon decays within some length scale as the irregularly spaced mass points can no longer sustain a standing wave: The phonon becomes a localized phonon. Formally, this is accounted for by setting wave vector q to be a complex quantity, $q = q` + iq``$ [11]. This modifies $u_n(t)$ to $\exp[i(q`an - \omega t)]$ $\exp(-q``an\omega t)$, hence, leading to an exponential decrease of amplitude $u_n(t)$ along the chain (relaxation): The greater the disorder in the substrate, the faster drops relaxation term $\exp(-q``an\omega t)$—evidently governed by the magnitude of $q``$. Hence, the greater the disorder of the substrate, the shorter the range of $u_n(t)$, or the stronger the localization of the phonons. Note that this takes place whether the material is a fractal or not.

Polydispersity exponent. Consider a site occupation number (probability, p) on a lattice. Near the percolation threshold (critical occupation), the cluster mass distribution $f(M)$ follows a power law, $f(M) \sim M^{-\tau}$, where τ is the polydispersity exponent (it should not be confused with a relaxation time). As mass M is conserved during cluster aggregation, we require $\int_0^{M_{max}} Mf(M)dM$ = constant. Hence, $\tau > 2$. In general, for mass fractals, condition $qR > 1$ prevails [see Eq. (3.8)]; hence the largest cluster mass, $M_{max} = M_z$ (z-average mass, Section 3.1.1), is conveniently related to cluster radius R under condition $qR > 1$.

We now take the average of the law for monodisperse scattering, $I(q) \sim q^{-d}$, over weight-average mass M_w (Section 3.1.1) in order to obtain the appropriate scattering relation for polydisperse gels. Integrating q over M_w, $q \sim \int q M_w f(M_w)\, dM_w$, yields

$$q_{M_w} \sim \int_0^{qR} M_w^2 M_w^{-\tau}\, dM_w \sim q^{3-\tau},\ \text{or}\ I(q) \Rightarrow I(q_{M_w}) \sim (q^{3-\tau})^{-d} \sim q^{-d(3-\tau)}$$

Note that for narrow polydispersities, $\tau \approx 2$, Eq. (3.8) is regained [12].

Random walk describes the probability $p(t,\mathbf{r})$ of finding a random variable at time t at position $\mathbf{r}$ with $p(t,\mathbf{r})$ given by Gaussian probability distribution $p(t,\mathbf{r}) = (4\pi K_0 t)^{-\alpha/2} \exp\{-\mathbf{r}^2/4K_0 t\}$ in Euclidian spaces of dimension $\alpha = E$ or on fractal substrates of fracton dimension $\alpha = \bar{d}$. The probability of the walker to return to a given point, say $\mathbf{r}_0 = 0$, after time interval t, is thus given by $p(t,0) \sim t^{-\alpha/2}$.

Roughness exponent. The scaling exponent of surface irregularity width w with L, $w(L) \sim L^\alpha$. Symbol α is current in the literature and should not confuse with its other uses (*see random walk,* above). To appreciate the relation between numerical values of exponent α, $0 < \alpha < 1$, and the visual roughness of a surface, it is convenient to solve width $w \sim L^\alpha$ for basis length L, writing $L \sim w^{1/\alpha}$. Clearly, for $\alpha \Rightarrow 0$, length $L = L_0$ would be much longer than length $L = L_1$ for $\alpha \Rightarrow 1$, at one and the same width w. Now assume we squeezed L_0 and L_1 to the same basis length L. Obviously, the resulting objects would show a much more jagged profile for length L_0 (exponent $\alpha \approx 0$) than for L_1 ($\alpha \approx 1$). Therefore, exponent $\alpha \approx 0$ engenders a very jagged-rough surface, whereas value $\alpha \approx 1$ represents a relatively very smooth surface. This outcome should be kept in mind as it is intuitively "strange."

  Exponent α can then be understood to quantify scaling of width w with surface length L under the following conditions: (a) Scaling is valid after a steady-state value of the variance of width w is attained [see Eq. (6.2)]. (b) Scaling under different α is compared at some common basis surface length L. (c) Scaling w with L probes $w(L)$ under condition $L \Rightarrow 0$, the microscopic limit. Hence, roughness exponent α describes how the root-mean-square width $\langle [w(L)]^2 \rangle^{1/2}$ varies from linear length (sample size) L down to smallest distances. In other words, α describes α *local* situation. Figure 6.3 demonstrates this on three surface profiles under $H = 0.7$ (persistent), 0.5 (perfectly random), and 0.3 (antipersistent).

Spectral dimension, or *fracton,* is exponent $\bar{d}$ in the power law of *spectral density* $\varphi(\Xi) \sim \Xi^{(\bar{d}-2)/2}$, where $\varphi(\Xi)$ is defined by $\varphi(\omega^2)d\omega^2$, the number of all squared angular frequencies $\Xi \equiv \omega^2$ between ω^2 and $\omega^2 + d\omega^2$. In turn, spectral density $\varphi(\Xi)$ is related to the density of vibrational states $\rho(\omega) = \sum_n \delta(\omega - \omega_n)$, the number of all frequencies between ω and $\omega + d\omega$, by $\rho(\omega) \sim \omega\varphi(\Xi)$. Consequently,

$$\rho(\omega) \sim \omega(\omega^2)^{(\bar{d}-2)/2} \sim \omega^{\bar{d}-1}$$

which is the relation most frequently used in the fracton literature.

The appearance of fracton $\bar{d}$ in anomalous diffusion *and* in the spectral density (or the density of vibrational states) of low-frequency vibrational modes of fractal substrates is appreciated by considering that anomalous (that is, say, *retarded*) translational diffusion and fracton modes (that is, *strictly localized* longitudinal displacement motions) are two sides of the same coin. Both aspects move under equation of motion $(\partial/\partial t)|X(t)\rangle = -H|X(t)\rangle$ and its formal solution $|X(t)\rangle = \exp[-H(t - t_0)]|X(t_0)\rangle$ with quantum mechanical Hamiltonian $H = [(ih/2\pi)^2/2m](\partial^2 X/\partial t^2)$ of general fluxes (see Section 4.1). Indeed, their representative quantities $|\cdots\rangle$, namely, $p(t,0)$, the probability of the random walker to return to its starting position after the time interval t, and $\varphi(\Xi)$, the number of low-frequency acoustic (longitudinal) eigenmodes of squared frequencies ω^2 between ω^2 and $\omega^2 + d\omega^2$, are related by a Laplace transform:

$$\int \exp(-\Xi t)\varphi(\Xi)d\Xi \equiv \mathcal{L}\{\varphi(\Xi)\} = p(t,\mathbf{r}_0)$$

Let us see how it works out: With $\varphi(\Xi) \sim \Xi^{-1+\bar{d}/2}$, we obtain indeed $\mathcal{L}\{\varphi(\Xi)\} \sim t^{-\bar{d}/2} \sim p(t,0)$ [13, 14].

NOTES AND REFERENCES

[1] E. Meyer and E.-G. Neumann, *Physical Acoustics,* translated by J. M. Taylor, Jr. (Academic, New York, 1972).

[2] K. D. Möller and W. G. Rothschild, *Far-Infrared Spectroscopy* (Wiley, New York, 1971), see Chapter 3. Most of the instrumentation described is outdated by now, but the text gives a general idea.

[3] R. Orbach, *Science* **231,** 814 (1986).

[4] P. Evesque, *Energy Migration: Theory,* in *The Fractal Approach to Heterogeneous Chemistry,* D. Avnir, Ed. (Wiley, Chichester, UK, 1989), p. 81.

[5] C. P. Lindsay and G. D. Patterson, *J. Chem. Phys.* **73,** 3348 (1980).

[6] W. G. Rothschild, M. Perrot, and J.-M. De Zen, *J. Chem. Phys.* **95,** 2072 (1991).

[7] Hierarchical relaxation: R. G. Palmer, D. Stein, E. S. Abrahams, and P. W. Anderson, *Phys. Rev. Lett.* **53,** 958 (1984); J. Klafter and M. F. Shlesinger, *Proc. Natl. Acad. Sci. USA* **83,** 848 (1986). Fractal dynamics of glasses: T. S. Chow, *Macromol.* **25,** 440 (1992). Anomalous relaxation in fractal structures: A. Plonka, *Ann. Rep. Progr. Chem.* **89** C, 37 (1992); S. Fujiwara and F. Yonezawa, *Phys. Rev. E* **51,** 2277 (1995).

[8] M. F. Shlesinger, *Ann. Rev. Phys. Chem.* **39,** 269 (1988)].

[9] J. Galambos, *The Asymptotic Theory of Extreme Order Statistics,* 2nd ed. (Krieger, Melbourne, FL, 1987), Section 3.12.

[10] B. B. Mandelbrot and J. W. van Ness, *SIAM Rev.* **10,** 422 (1968).

[11] See [2], pp. 413–415.

[12] For a more precise presentation, see J. E. Martin and B. J. Ackerson, *Phys. Rev. A* **31,** 1180 (1985).

[13] I. S. Gradshsteyn and I. M. Ryzhik, *Table of Integrals, Series, and Products* (Academic, New York, 1965), see entry 3.381.4.

[14] For a clear and concise derivation of the (entirely general) relation between $p(t,0)$ and $\varphi(\Xi)$, refer to P. Pfeifer and M. Obert, *Fractals: Basic Concepts and Terminology,* in *The Fractal Approach to Heterogeneous Chemistry*, D. Avnir, Ed. (Wiley, Chichester, UK, 1989), p. 28. However, it may be helpful to first refresh one's mind in P. A. M. Dirac, *The Principles of Quantum Mechanics* (Oxford University Press, London, 1958), about $\langle$bra$|$ $|$ket$\rangle$ vector notation, the definitions of the delta function $\int f(x)\delta(x-a)dx = f(a)$, about the average value of a matrix element $\langle b|a\rangle = \int \langle b|\xi'\rangle\delta(\xi'-\xi'')\langle\xi''|a\rangle d\xi''$ of states $|a\rangle, |b\rangle$, and about complete sets of orthonormal basis vectors $|\xi'\rangle$.

AUTHOR INDEX

Chapter 7 (Footnote number; Section number)

SUBJECT INDEX

Ablation processes, of
    instant coffee particles by mechanical
      tapping, fractal perimeter
      dimensions of, 116–119
    polymers by laser irradiation, surface
      fractal dimensions of, 164
Activated charcoal
    nitrogen isotherms on, 22
    pore dimension in the presence of
      hydrostatic forces, 22
Adjacency relations
    and connectivity of a fractal structure,
      79, 90
    fracton characterization of, 102–104,
      109–111
    in connected structures of water,
      102–104
    in electron transfer of silica-
      adsorbed anthracene radical cation,
      110–111
    measurement of, 177
    metric (Manhattan) in, 136–137,
      194
    topological indexes of, 133

Adsorbate(s)
    adsorbate–adsorbate interactions,
      detrimental effects on surface
      analysis, 15
    by homologous series of molecules,
      in adsorbent surface tiling method,
      13, 24
    as yardstick measure in surface tiling,
      6, 29–32
Adsorbate–adsorbent interactions
    through hydrostatic forces, 20–24
    minimization of, by use of nitrogen
      as adsorbate, 33
    neglect of, effect on adsorbent
      surface tiling, 15
    relative strength in physisorption and
      chemisorption, 12
    in size-exclusion gel chromatography,
      perturbing effects on, 49
Adsorbate uptake
    and adsorbent self-similarity, 16
    of chain molecules, nonequilibrium
      state and steric hindrance of,
      32–33

degree of, with surface fractal
dimension
on gamma-alumina, 13–14
on preadsorbed films of water and
n-heptane, 26–28
of temperature-shocked gibbsite,
28–30
by zeolites, fractal aspects of extra-
frame surface sites in, 24–25
Adsorbent
fractal dimension and space-filling
capacity, 13–14
generation of effective surface
dimension of, by preloaded
molecular films, 28
self-similarity and principle of the
inverse tiling method, 17–19
Adsorption and catalysis, effects of
surface in, 2
Aerogels, silica
fractal mass/surface dimensions from
small-angle neutron scattering, 108
fracton dimension
from Brillouin scattering, 109–110
from inelastic neutron scattering,
109
properties of, 107–108
Agarose (a sugar), gel agent for
convection-damping during
electrolysis, 63
Aggregation
crossover between diffusion-limited
and reaction-limited processes,
68–71
by diffusion-limited cluster–cluster
processes, 51–53
by diffusion-limited particle–cluster
processes, 50–51
flocculation mechanisms, of silica,
106
fractality and general principles of,
46–47
by reaction-limited cluster–cluster
processes, 68–69, 106

space dimension effects in, 50
space-filling capacity and fractal
character of, 71
Alkanes
adsorption by temperature-treated
gibbsite, 27–28
molecular flexibility and boiling
points
of monohalogenated and
perfluorinated, 137–138
principles of scaling method,
134–135
sorption kinetics of
in urethane-modified bismaleimide
elastomers, 153–154
in vulcanized rubbers, 148–153
topological indexes (graph invariants)
for, 133–137
Alpha (power law exponent α)
in random walk, probability to return
to origin, 204
as surface roughness exponent
derivation by scaling arguments,
129, 204
relation to
exponent *H* of fractional
Brownian motion, 165
surface fractality, 164–165
in silver electrodeposition, 122
of water–paper towel interface, 155
Aluminum oxide (gamma-alumina)
dimension from tiling by coiled
polymer chains, 13–14
discrimination against uptake of
larger adsorbate molecules, 30
dispersive agent for rare metals,
38–40
fractal dimension of, 13
generation from temperature-shocked
gibbsite, 28–30
as high-area fractal surface/pore
adsorbent, 12–14
Anomalous diffusion
definition and principles, 77–78, 199